## Mechanical Engineering Series

Frederick F. Ling
*Series Editor*

Mark S. Darlow
Department of Mechanical Engineering, Rensselaer Polytechnic Institute, Troy, New York 12180-3590, USA

*Series Editor*

Frederick F. Ling
Director, Columbia Engineering Productivity Center and Professor, Department of Mechanical Engineering, Columbia University, New York, New York 10027-6699;

William Howard Hart
Professor Emeritus, Department of Mechanical Engineering, Aeronautical Engineering and Mechanics, Rensselaer Polytechnic Institute, Troy, New York 12180-3590, USA

Darlow, M.
Balancing of high-speed machinery / Mark S. Darlow.
p. cm.
Bibliography: p.
Includes index.
ISBN 0-387-96986-1 (alk. paper)
1. Balancing of machinery. 2. Turbomachines—Vibration.
I. Title.
TJ177.D37 1989
621.8′11—dc19 89-5992

Printed on acid-free paper.



Camera-ready copy prepared by the author using Microsoft Word.
Printed and bound by R. R. Donnelley & Sons, Harrisonburg, Virginia.
Printed in the United States of America.

9 8 7 6 5 4 3 2 1

ISBN 0-387-96986-1 Springer-Verlag New York Berlin Heidelberg
ISBN 3-540-96986-1 Springer-Verlag Berlin Heidelberg New York

# Series Preface

Mechanical engineering, an engineering discipline borne of the needs of the industrial revolution, is once again asked to do its substantial share in the call for industrial renewal. The general call is urgent as we face profound issues of productivity and competitiveness that require engineering solutions, among others. The Mechanical Engineering Series is a new series, featuring graduate texts and research monographs, intended to address the need for information in contemporary areas of mechanical engineering.

The series is conceived as a comprehensive one which will cover a broad range of concentrations important to mechanical engineering graduate education and research. We are fortunate to have a distinguished roster of consulting editors, each an expert in one of the areas of concentration. The names of the consulting editors are listed on the first page of the volume. The areas of concentration are: applied mechanics; biomechanics, computational mechanics; dynamic systems and control; energetics; mechanics of materials; processing; thermal science; and tribology.

Professor Marshek, the consulting editor for dynamic systems and control, and I are pleased to present the second volume of the series: *Balancing of High-Speed Machinery* by Professor Darlow. The selection of this volume underscores the interest of the Mechanical Engineering Series to provide our readers with topical monographs as well as graduate texts.

Frederick F. Ling

# Preface

Modern rotating machinery, particularly turbomachinery, is frequently being designed to operate at higher speeds than in the past. Consequently, there is an increased need to balance high-speed rotors; and thus the motivation for this book.

There are important distinctions between balancing high-speed and low-speed rotors. The balancing of low-speed, or rigid, rotors is a relatively straightforward process for which a variety of machines are commercially available. These machines provide an effective and efficient means for balancing such rotors; but are much less effective for balancing high-speed rotors. Often, individual components of high-speed rotors are balanced separately using low-speed balancing machines, prior to assembly. However, if the mounting of these components is eccentric with respect to the center of rotation, which is invariable true to a greater or lesser extent, the effectiveness of the component balancing will be compromised. Thus, while such an approach may serve to simplify the balancing of the assembled high-speed rotor, it does not usually eliminate the need for balancing the fully assembled rotor.

Rotor balancing is necessary due to the presence of eccentricity of the mass centroidal axis of a rotating shaft, relative to the center of rotation. Such eccentricity generates centrifugal forces whose amplitudes are proportional to the square of the rotational speed. Rotors which operate at high speeds tend to be more slender, and thus more flexible, than those that operate at low speeds. Consequently, high-speed rotors often deform significantly under the action of these unbalance forces. Rotors which exhibit such behavior are generally referred to as *flexible rotors*.

It is not unusual for the maximum operating speed of a flexible rotor to be above several synchronous resonances, or critical speeds. When the operating speed, even temporarily, is in the vicinity of one of these critical speeds, there is a magnification of the response due to the mass eccentricity and the flexible rotor deforms in a shape which is most strongly influenced by that particular resonance. Thus, the deformed shape of the rotor changes dramatically with changes in the rotational speed, as does the effect of the mass unbalance distribution. Therein lies the difficulty of balancing flexible rotors. Since low-speed balancing machines operate on the premise that the effect of the mass unbalance distribution is not speed-dependent, other approaches must be used to balance flexible rotors. Such approaches are generally referred to as *flexible rotor balancing*, or as *high-speed rotor balancing*.

Most flexible rotor balancing practices can be classified into three groups: modal balancing, influence coefficient balancing and the Unified Balancing Approach. Modal balancing methods assume orthogonal modes of vibrational deformation and strive to compensate for them individually. Influence coefficient balancing methods determine the effect of a unit unbalance mass at various stations along the longitudinal axis of the rotor and select a set of discrete correction masses that will minimize vibrational amplitudes at other selected locations along this axis. The Unified Balancing Approach applies the empirical nature of influence coefficient balancing while taking advantage of the modal behavior of the rotor and, thus, seeks to combine the best features of both modal and influence coefficient balancing.

The purpose of this book is to provide the engineering student or practicing engineer with a single, complete reference on high-speed rotor balancing. To this end, detailed analytical background and practical application procedures are presented for each of the principal high-speed rotor balancing methods. This information is supplemented and supported through a presentation of the theoretical development of synchronous rotor vibration and a brief overview of rigid rotor balancing techniques and machines. While most of the material in this book can be found in the technical literature, it has not previously been available in a single, concise volume, together with detailed descriptions of application procedures.

## Acknowledgements

The author would like to thank the many people who have contributed, both directly and indirectly, to the publication of this book. The knowledge gained from his colleagues at Mechanical Technology Incorporated, the University of Florida and the Rensselaer Polytechnic Institute have made the authoring of this volume possible. The book was completed while the author served as *Fohrman Visiting Professor* of Mechanical Engineering at the Technion - Israel Institute of Technology. Ms. Hollis McEvilly is responsible for most of the typing and editing that went into this book, as well as the handling of all details that could not be taken care of from six thousand miles away. Her tireless efforts are greatly appreciated. The author would also like to thank the publisher, Springer-Verlag.

Finally, I would like to thank my family for providing both moral and practical support. My wife, Myra, invested many hours of typing, editing and proofreading, and a great deal of patience in the book. My older son, Adam, was responsible for a number of the figures; and my younger son, Eric, was willing to give up the computer when needed to write this book.

December 1988 *Mark S. Darlow*

# Contents

# LIST OF SYMBOLS

## Roman Symbology

| | |
|---|---|
| $A$ | Cross-sectional area of rotor |
| $\hat{\mathbf{A}}$ | Complex influence coefficient matrix |
| $a$ | Pure radial mass eccentricity of rotor |
| $\hat{a}(z)$ | Total rotor complex-valued mass eccentricity |
| $\hat{a}_i$ | Complex modal mass eccentricity for $i^{\text{th}}$ mode |
| $a_u(z)$, $a_v(z)$ | Components of $\hat{a}(z)$ in rotating coordinate system |
| $a_{iu}$, $a_{iv}$ | Components of $\hat{a}_i$ in rotating coordinate system |
| $\hat{\mathbf{a}}_i$ | $i^{\text{th}}$ column of influence coefficient matrix |
| $b$ | Viscous damping coefficient |
| $E$ | Modulus of elasticity |
| $\mathbf{E}_k$ | Influence coefficient iterative weighting matrix for $k^{\text{th}}$ iteration |
| $e$: | Transcendental number equal to 2.71828182 . . .[also exp()] |
| $\hat{e}_i$ | $i^{\text{th}}$ complex-valued basis vector |
| $\mathbf{F}$ | Influence coefficient general weighting matrix |
| $f_i$ | $i^{\text{th}}$ diagonal element of $\mathbf{F}$ |
| $I$ | Cross-sectional area moment of inertia; also used for $(I_p - I_t)$ |
| $I_p$ | Polar (gyroscopic) mass moment of inertia |
| $I_t$ | Transverse (rotatory) mass moment of inertia |
| $i$ | Square root of negative one |
| $k$ | Linear stiffness coefficient (spring rate) |
| $l$ | Generally used to indicate rotor length |
| $m$ | Mass |
| $m_i$ | $i^{\text{th}}$ added mass for balancing |
| $O$ | Origin of stationary and rotating coordinate systems |

| | |
|---|---|
| $Q$ | Modal normalizing factor |
| $\hat{R}$ | Complex value representing rotor vibration in a rotating coordinate system |
| $r$ | Magnitude of $\hat{R}$ |
| $\hat{r}_i$: | Complex value representing radius and angular location of $m_i$ |
| $S_f$ | Significance factor used for balancing plane optimization |
| $t$: | Time |
| $U$ | Abscissa in rotating coordinate system |
| $u$ | Component of vibration vector along the $U$ axis |
| $\hat{u}$ | Complex-valued rotor mass imbalance |
| $u_c$, $u_s$ | Components of $\hat{u}$ in rotating coordinate system |
| $V$ | Ordinate in rotating coordinate system |
| $v$ | Component of vibration vector along the $V$ axis |
| $\hat{\mathbf{w}}$ | Vector of complex-valued trial or correction masses |
| $\hat{w}_i$ | $i^{th}$ component of $\hat{\mathbf{w}}$ |
| $X$ | Abscissa in stationary coordinate system |
| $x$ | Component of vibration vector along the $X$ axis |
| $\hat{x}$ | Complex-valued rotor vibration amplitude |
| $x_c$, $x_s$ | Components of $\hat{x}$ in stationary coordinate system |
| $\hat{\mathbf{x}}$ | Vector of complex vibration values |
| $\hat{\mathbf{x}}_w$ | Vibration vector due to mass set $\hat{\mathbf{w}}$ |
| $Y$ | Ordinate in stationary coordinate system |
| $y$ | Component of vibration vector along the $Y$ axis |
| $Z$ | Coordinate lying along the rotor axis |
| $z$ | Distance from origin along the rotor axis |

## Greek Symbology

| | |
|---|---|
| $\alpha_{ij}$ | $ij^{th}$ element of influence coefficient matrix |
| $\beta$ | Lagging phase angle in Jeffcott analysis |
| $\beta_\theta$ | Lagging phase angle in conical Jeffcott analysis |
| $\varepsilon_{ik}$ | $i^{th}$ influence coefficient weighting factor for $k^{th}$ iteration |
| $\zeta$ | Damping ratio for unidirectional vibration or Jeffcott analysis |
| $\zeta_i$ | Damping ratio for $i^{th}$ mode in modal analysis |

| | |
|---|---|
| $\zeta_\theta$ | Damping ratio for conical Jeffcott analysis |
| $\hat{\eta}(z)$ | Total rotor whirl deformation |
| $\hat{\eta}_i$ | Rotor whirl deformation for $i^{th}$ mode |
| $\eta_u(z)$, $\eta_v(z)$ | Components of $\hat{\eta}(z)$ in rotating coordinate system |
| $\eta_{iu}$, $\eta_{iv}$ | Components of $\hat{\eta}_i$ in rotating coordinate system |
| $\theta$ | Angular dynamic displacement of rotor |
| $\theta_a$: | Pure angular mass eccentricity of rotor |
| $\lambda_i$ | $i^{th}$ rotor system eigenvalue |
| $\Pi$ | Indicates mathematical multiplication series |
| $\pi$ | Standard constant 3.14159265 . . . |
| $\rho$ | Mass density |
| $\Sigma$ | Indicates mathematical summation series |
| $\phi$ | Leading phase angle |
| $\phi_i(z)$ | $i^{th}$ rotor system eigenvector (mode shape) |
| $\omega$ | Rotational frequency of rotor |
| $\omega_{cr}$ | Rotor critical speed for Jeffcott analysis |
| $\omega_i$ | Undamped critical speed for $i^{th}$ mode in modal analysis |
| $\omega_n$ | Undamped natural frequency |
| $\omega_{n\theta}$ | Undamped rotatory natural frequency |

## Explanation of Notation

| | |
|---|---|
| $\dot{f}$ | Indicates differentiation with respect to time |
| $\hat{f}$ | Indicates a complex quantity (often dropped for convenience where no ambiguity exists) |
| $\mathbf{f}$ | Indicates (real or complex) vector quantity |
| $\mathbf{F}$ | Indicates (real or complex) matrix quantity |
| $f_c$, $f_s$ | Indicates components in rotating coordinate system |
| $f_x$, $f_y$ | Indicates components in stationary coordinate system |

Additional nomenclature which is not used consistently throughout the book is defined in the context in which it is used.

# Chapter 1
# Introduction

The mass balancing of high-speed rotors is an integral part of the study and practice of rotordynamics. As is implied by the name, rotordynamics is concerned with the dynamics of rotating machinery. In this context, dynamics refers to harmonic motion, or vibration. Any machine with rotating components is considered to be a rotating machine. The great majority of commercial machinery falls into this category. Rotating machines range in size from small gyroscopes weighing only a few ounces to large rock tumblers weighing several tons. Other examples of rotating machines are household electric motors, internal combustion engines, gas turbine engines, steam turbines, electric generators, industrial compressors and power transmission systems, to name just a few.

The part of a rotating machine that rotates in normal operation is generally referred to as the rotor. The part that does not rotate is referred to as the stator. The rotor is flexible if it deforms when the machine is operating at any speed up to its maximum operating speed, including designed overspeed conditions. Otherwise, the rotor is rigid. Of principal interest here are machines with flexible rotors.

The general purpose of the study of rotordynamics is to increase understanding of rotor vibration phenomena and thus provide a means for controlling or eliminating these vibrations. As the efficiency of rotating machinery has been increased through reduced weight and increased speed, which implies increased rotor flexibility, the control of rotor vibrations has become essential. Noise reduction was at one time the primary motivation for the control of rotor vibrations. However, for current machinery designs, control of rotor vibration is necessary to ensure machinery survival and operator safety. In the case of commercial aircraft, the safety of hundreds of passengers, per flight, is the primary consideration. Thus, the study and practice of rotordynamics has taken on an increasingly important role in recent years.

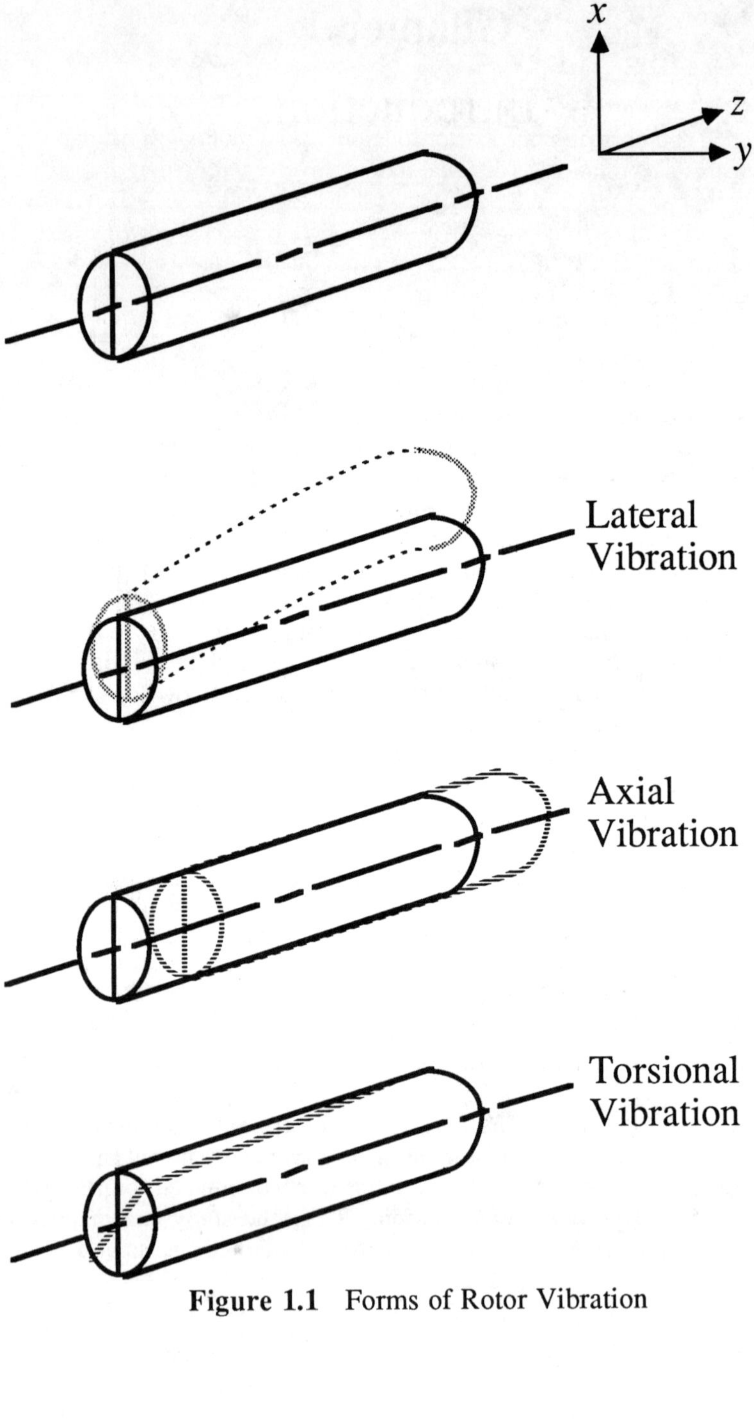


**Figure 1.1** Forms of Rotor Vibration

Rotor vibration is generally classified according to the direction and frequency of the vibratory motion. The three possible directions of motion, illustrated in Figure 1.1, are lateral, axial, and torsional. Lateral motion involves a displacement from its central position, flexural deformation and/or rotation of the rotor's axis of rotation. In this context, rotation is about an axis intersecting and normal to the axis of rotation. Axial motion occurs parallel to the rotor's axis of rotation. Torsional motion involves a rotation of the rotor's transverse sections relative to one another about its axis of rotation. Vibrations which occur at the frequency of rotation of the rotor are referred to as synchronous vibrations. Those which occur at other frequencies are known as nonsynchronous vibrations. In geared systems, where two or more rotors are operating at different speeds, a different synchronous frequency is associated with each rotor. Synchronous lateral vibration is a frequent cause of machine failure and probably the most common source of machine noise and vibration. Rotating mass unbalance results only in synchronous lateral vibration and is its most common cause. Thus, other forms of rotor vibration are of little interest in a study of rotor balancing, such as is presented here. Consequently, further reference to rotor vibration herein generally implies synchronous lateral vibration.

The presence of rotor mass imbalance, more commonly known as *unbalance*, is due to the eccentricity of the mass-centroidal axis of the rotor relative to its axis of rotation. In practice, this unbalance is generally a result of unavoidable imperfections in rotor manufacture and assembly. The term *rotor balancing* covers a broad range of procedures aimed at reducing rotor unbalance. The ultimate purpose of all rotor balancing procedures is to produce a smooth-running machine. This is achieved in practice not by removing all unbalance, but rather by applying compensating unbalances, or by removing mass by machining, which result in a smooth-running machine. Since rotor unbalance is generally distributed along the length of the rotor and compensating unbalances are applied at discrete axial locations, the complete elimination of unbalance is usually impossible, as well as unnecessary.

Rotor balancing procedures are classified according to the type of rotor for which they are designed: rigid or flexible. When balancing rigid rotors, the shape of the centroidal axis does not change as a function of speed. For flexible rotors, the shape of the centroidal axis, and thus the unbalance distribution, does change with speed. Thus, in general, rigid rotor balancing procedures and tools are not applicable to the balancing of flexible rotors. The procedures for balancing rigid rotors are well established and accepted. However, there is much less agreement on flexible rotor balancing procedures.

Most flexible rotor balancing can be classified into two groups, modal balancing and influence coefficient balancing. Modal balancing methods assume orthogonal modes of vibrational deformation and strive to compensate for them. Influence coefficient balancing methods determine the effect of a unit unbalance mass at various stations along the z axis and select a set of discrete correction masses that will minimize vibrational amplitudes at other selected locations along the z axis. A third method, the Unified Balancing Approach, is, in a sense, a combination of modal and influence coefficient balancing. It applies the empirical nature of influence coefficient balancing while taking advantage of the modal behavior of the rotor. This book provides detailed descriptions of these flexible rotor balancing methods including analytical bases, application procedures and practical results.

The following topics are discussed in subsequent chapters: (1) a basic review of vibration and synchronous rotordynamic theory; (2) a review of the literature on rotor balancing with emphasis on flexible rotor balancing; (3) a detailed review and discussion of rigid rotor balancing and flexible rotor balancing procedures; (4) an enhancement of the influence coefficient balancing method (balancing plane optimization procedure) designed to eliminate the possibility of using redundant balancing planes; (5) results of actual application of flexible rotor balancing procedures; and (6) the use of the principle of reciprocity to reduce the number of balancing runs. A glossary of terms is also included along with an extensive list of references.

# Chapter 2
# Theoretical Background

In recent years, the trend in the design of rotating machinery has been toward reduced weight and increased operating speeds, with the objective of increasing operating efficiency and thus reducing cost. However, these more efficient designs result in increased rotor flexibility and are, in general, more susceptible to a variety of undesirable rotordynamic phenomena. In particular, increased rotor flexibility substantially complicates rotor balancing requirements. Control of rotor vibration is necessary to maintain reasonable noise levels and to ensure operator, and consumer safety, and machinery survival. In general, the operating expense of rotating machinery is a direct function of the success with which rotor vibrations are controlled.

Rotor vibration is generally classified into two groups. The first group is synchronous vibration which is usually controlled through rotor balancing. The second group is nonsynchronous vibration, which is often unstable and is usually controlled through the addition of external damping. The former of these groups is of interest here.

To fully appreciate the rotor balancing problem, an understanding of basic vibration theory is necessary. A brief overview of unidirectional vibration theory is presented in this chapter for this purpose. Many references are available which provide a much more complete presentation of vibration theory, among them references [38,134,135,139].

An introduction to the theory of rotor-bearing vibrations is also presented in this chapter, with emphasis on synchronous vibration. Relatively few references are available which provide a complete treatment of rotor-bearing vibration. The most prominent of these are references [40,83,116,136].

## Basic Vibration Theory

A vibratory system may be considered to be composed of a number of individual elements, each with the capability to store potential and kinetic energy and to dissipate energy. The vibration phenomenon is a manifestation of an alternating transfer of energy between kinetic and potential. For simplicity, system representations are generally used whose elements each have only one of the three capabilities mentioned above. These representations are referred to as lumped parameter models. The number of these elements required to accurately represent a real vibratory system depends on a number of factors including the geometry and loading of the system and the anticipated excitation frequencies.

For our purposes, these lumped parameter model elements are assumed to behave linearly. This is, in general, a reasonable assumption for small vibratory motions, although it may not be as accurate for large motions. In the rotor balancing problem, the small motion assumption is rarely violated. In any case, most balancing procedures can tolerate modest system nonlinearities, and many of these procedures, through the use of repeated balancing, can efficiently handle rotor-bearing systems with substantial non-linearity.

The three lumped parameter model elements are: 1) the spring for storing potential energy, 2) the mass for storing kinetic energy and 3) the damper for dissipating energy. A linear spring exerts a force proportional to its change in length and in such a direction as to attempt to restore the spring to its original length. This type of force is commonly referred to as a restoring force. The proportionality constant between the deflection of the spring and the restoring force is alternatively known as spring stiffness, spring constant or spring rate, and is represented by the letter $k$. This ideal, lumped-parameter spring is massless so that the forces at the ends of the spring are equal in magnitude and opposite in direction.

For a mass, which stores kinetic energy, the force is proportional to acceleration. The proportionality constant here is mass (as distinguished from weight) and is represented by the letter $m$. A lumped-mass element, used for modeling purposes, is assumed to be rigid.

Of several possible representations for an energy dissipation element, the viscous damper is most commonly used. A discussion of other damper models is beyond the scope of this book. The restoring force for a viscous damper is proportional to the relative velocity between the attachment points, or ends, of the damper. The proportionality constant, referred to as the viscous damping constant or coefficient, is represented by the letter $b$. The ideal damper, like the ideal spring, is massless so that the forces at the ends are equal in magnitude and opposite in direction.

## Undamped Free Vibration Model

The simplest possible vibratory system representation is a single-degree-of-freedom model for free vibration with no damping, as illustrated schematically in Figure 2.1. The term "free vibration" indicates that the system is provided with a set of initial conditions and then allowed to vibrate freely, with no external influences. For the model in Figure 2.1, a mass is attached to a rigid support by a linear spring. The motion of the mass is unidirectional with displacement $x$, where $x$ is a function of time. The differential governing equation for this system may be obtained through the use of Newton's second law of motion, which may be stated as, "the rate of change of momentum is equal to the vector sum of forces exerted on the body and takes place in the direction in which the force (vector sum) acts". If the mass is constant, the rate of change of momentum is equal to the mass times its acceleration.

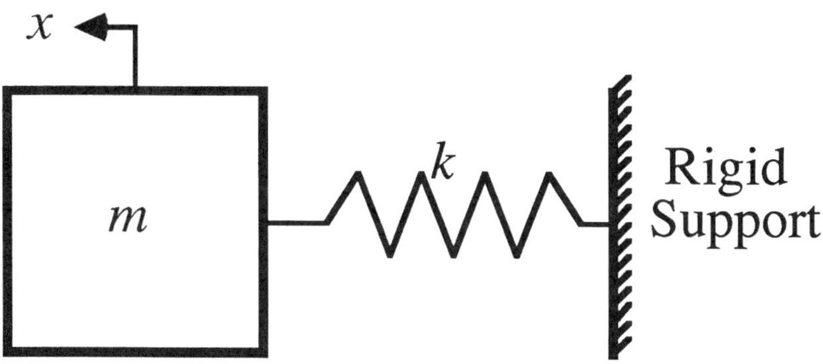


**Figure 2.1** Single Degree-of-Freedom, Undamped, Free Vibration Model

Since the system considered here is unidirectional, this law reduces to

$$m\ddot{x} = \Sigma \text{ (forces in the } X\text{-direction)} = -kx \tag{2.1}$$

where the dots over the $x$ represent derivatives with respect to time and the negative sign indicates that the spring force is a restoring force. The form of the restoring force indicates that the static equilibrium position is assumed to be at the point where $x$ equals zero.

Rearrangement of Equation (2.1) results in the more familiar form of the governing equation given by

$$m\ddot{x} + kx = 0 \tag{2.2}$$

The solution to this differential equation is given by

$$x = A \sin \omega_n t + B \cos \omega_n t \tag{2.3}$$

where the constants $A$ and $B$ are determined from the initial conditions of the system and $\omega_n$, known as the natural frequency or resonance, is given by

$$\omega_n = \sqrt{k/m} \tag{2.4}$$

The units of $\omega_n$ are radians per second.

Equation (2.3) may be rewritten in the form

$$x = C \sin(\omega_n t + \theta) \tag{2.5}$$

where $C$ is the maximum amplitude of vibration (single amplitude) given by

$$C = \sqrt{A^2 + B^2} \tag{2.6}$$

and

$$\theta = \arctan(B/A) \tag{2.7}$$

The phase angle, $\theta$, is an indication of the point in the vibration cycle at which the value of $x$ becomes equal to zero. The term *phase angle*, when used in different contexts, can take on different meanings. This form of vibration is referred to as simple harmonic motion due to the harmonic, or cyclic, form of Equations (2.3) and (2.5).

## Damped Free Vibration Model

The vibratory system of Figure 2.1 is made slightly more complicated by the addition of a viscous damping element as shown in Figure 2.2. Again considering only free vibration, the differential governing equation can be derived from Newton's second law to give

$$m\ddot{x} + b\dot{x} + kx = 0 \tag{2.8}$$

where $b$ is the viscous damping coefficient and the other parameters are as defined above. The form of the solution to this differential equation depends on the value of $b$ relative to the values of $m$ and $k$, assuming all are positive. The criteria for the selection of the appropriate solution is whether the damping coefficient, $b$, is less than, equal to or greater than the critical damping value, $b_c$, which is defined by

$$b_c = 2\sqrt{km} = 2m\omega_n \tag{2.9}$$

where $\omega_n$ is defined as before to be $\sqrt{k/m}$. The ratio of the damping coefficient to the critical damping is referred to as the damping ratio, $\zeta$.

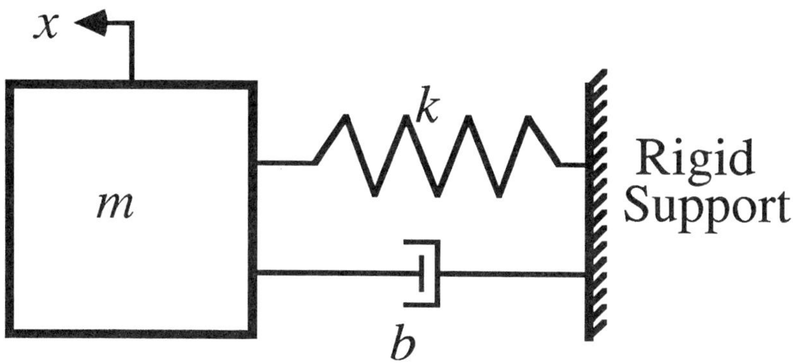


**Figure 2.2** Single Degree-of-Freedom Free Vibration Model with Viscous Damping

For the case in which the damping is less than critical, or underdamped ($\zeta < 1$), the solution to Equation (2.8) is of the form

$$x = e^{-bt/2m}\left(A \sin \omega_d t + B \cos \omega_d t\right) = Ce^{-bt/2m} \sin(\omega_d t + \theta) \tag{2.10}$$

where $C$ and $\theta$ are related to $A$ and $B$ by Equations (2.6) and (2.7). $\omega_d$ is called the damped natural frequency and is given by

$$\omega_d = \sqrt{k/m}\left(1 - \zeta^2\right)^{1/2} = \omega_n \left(1 - \zeta^2\right)^{1/2} \tag{2.11}$$

in units of radians per second. Note that for unidirectional vibration, the damped natural frequency, $\omega_d$, is less than the undamped natural frequency, $\omega_n$. The vibratory response represented by Equation (2.10) is referred to as a damped oscillation. A plot of typical damped oscillation response is presented in Figure 2.3.

For the case in which the damping is exactly critical, or critically damped ($\zeta = 1$), the solution to Equation (2.8) is of the form

$$x = (A + Bt)e^{-bt/2m} = (A + Bt)e^{-\omega_n t} \tag{2.12}$$

where $A$ and $B$ are constants which depend on the initial conditions. For this case, a typical response plot is presented in Figure 2.4.

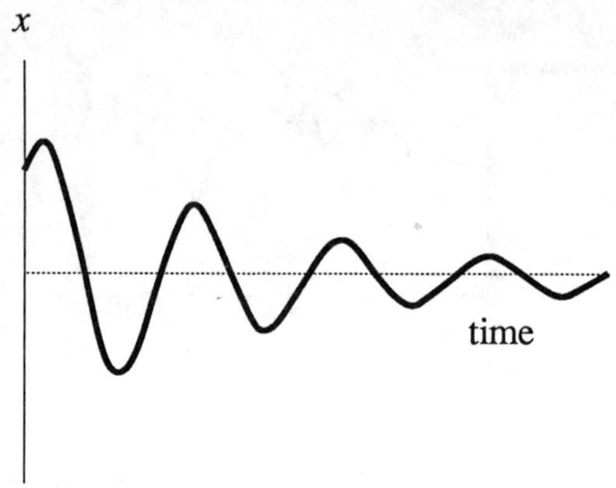


**Figure 2.3** Typical Free Vibration Response for a Single Degree-of-Freedom System with ($\zeta < 1$)

For the case in which the damping is greater than critical, or overdamped ($\zeta > 1$), the solution to Equation (2.8) is of the form

$$x = \exp(-bt/2m)\left[A\exp\left(\omega_n t\sqrt{\zeta^2 - 1}\right) + B\exp\left(-\omega_n t\sqrt{\zeta^2 - 1}\right)\right] \tag{2.13}$$

where $A$ and $B$, as before, are constants which depend on the initial conditions. A typical response plot for this case is presented in Figure 2.5.

Most practical vibratory systems are underdamped ($\zeta < 1$). Critically damped and overdamped systems, in general, do not exhibit rotor resonance problems. However, the presence of rotor mass unbalance in such systems can result in excessive bearing forces at high speeds. Thus, even overdamped systems often require rotor balancing. However, since this is a relatively rare situation, and in such cases rigid rotor balancing is generally sufficient, the remainder of this book considers only underdamped systems.

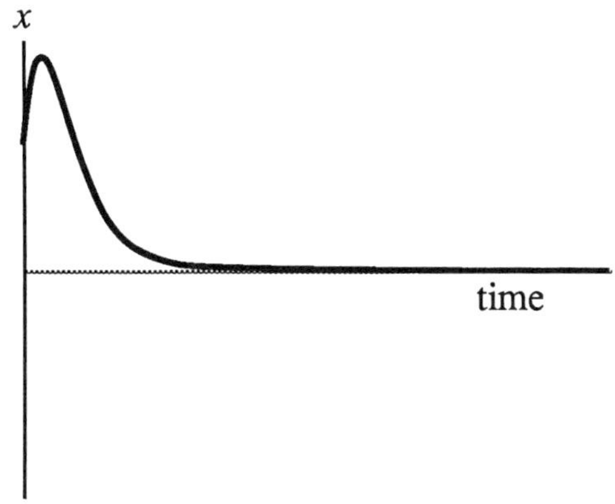


**Figure 2.4** Typical Free Vibration Response for a Single Degree-of-Freedom System with ($\zeta = 1$)

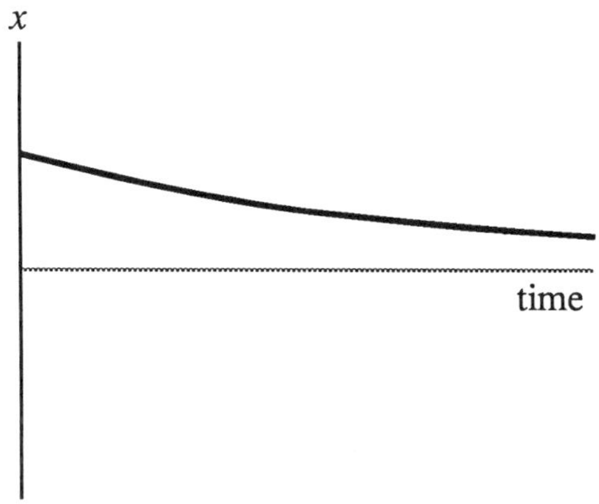


**Figure 2.5** Typical Free Vibration Response for a Single Degree-of-Freedom System with ($\zeta > 1$)

## Undamped Forced Vibration Model

Forced vibration is now reviewed, beginning with a single degree-of-freedom undamped system, as illustrated in Figure 2.6. The forcing function is assumed to be sinusoidal and may be realized as a force applied to the mass, as shown in Figure 2.6a, or as a prescribed motion of the foundation, as shown in Figure 2.6b. The response of the system is defined by the force or motion transmissibility, depending on the form of the forcing function. The force transmissibility is defined as the ratio of the force transmitted to the foundation to the force applied to the mass. The motion transmissibility is defined as the ratio of the motion of the mass to the applied motion of the foundation.

For the case in which a sinusoidal forcing function is applied to the mass, the governing equation becomes

$$m\ddot{x} + kx = F_o \sin \omega t \tag{2.14}$$

where $F_o \sin \omega t$ is the forcing function. Since Equation (2.14) is not homogeneous, it has both a general and a particular solution which together are given by

$$x = A \sin \omega_n t + B \cos \omega_n t + \left( \frac{F_o / k}{1 - \omega^2 / \omega_n^2} \right) \sin \omega t \tag{2.15}$$

where $A$ and $B$ are constants that depend on the initial conditions and $\omega_n$ is, as before, equal to $\sqrt{k/m}$. The first two terms on the right side of Equation (2.15) form the general solution to Equation (2.14). This is identical to the solution for the analogous free vibration case, given by Equation (2.3). In virtually all practical systems there is some damping present which causes the general solution terms (also known as the transient solution) to disappear, or damp out, with time. However, the particular, or steady-state, solution given by the last term in Equation (2.15) does not damp out and is, therefore, of primary interest here. This steady-state solution is analogous to synchronous vibration in rotating machinery, as will be shown below.

Considering only the steady-state solution, the force transmissibility, $T_F$, is given by

$$T_F = \frac{kx}{F_o \sin \omega t} = \left( 1 - \frac{\omega^2}{\omega_n^2} \right)^{-1} \tag{2.16}$$

where $T_F$ is unbounded when the forcing frequency, $\omega$, is equal to the natural frequency, $\omega_n$. A plot of $T_F$ as a function of the frequency ratio, $\omega / \omega_n$, is presented in Figure 2.7.

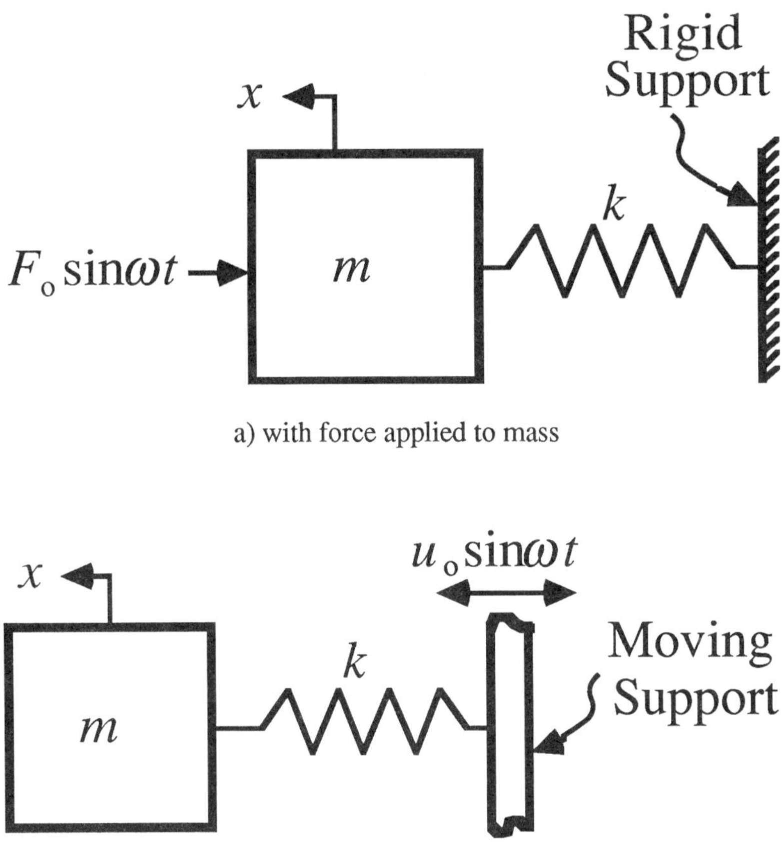


a) with force applied to mass



b) with prescribed motion of foundation

**Figure 2.6** Single Degree-of-Freedom, Undamped, Forced Vibration Model

For the case in which the foundation has a prescribed sinusoidal motion, the governing equation becomes

$$m\ddot{x} + kx = ku_o \sin \omega t \tag{2.17}$$

where $u_o \sin\omega t$ is the prescribed motion of the support. Equation (2.17) has a solution very similar to Equation (2.15) given by

$$x = A \sin \omega_n t + B \cos \omega_n t + \left( \frac{u_o}{1 - \omega^2 / \omega_n^2} \right) \sin \omega t \tag{2.18}$$

where $A$ and $B$ are constants that depend on the initial conditions and $\omega_n$ is equal to $\sqrt{k/m}$. As in Equation (2.15), the first two terms on the right side of Equation (2.18) form the transient solution, while the last term represents the steady-state solution.

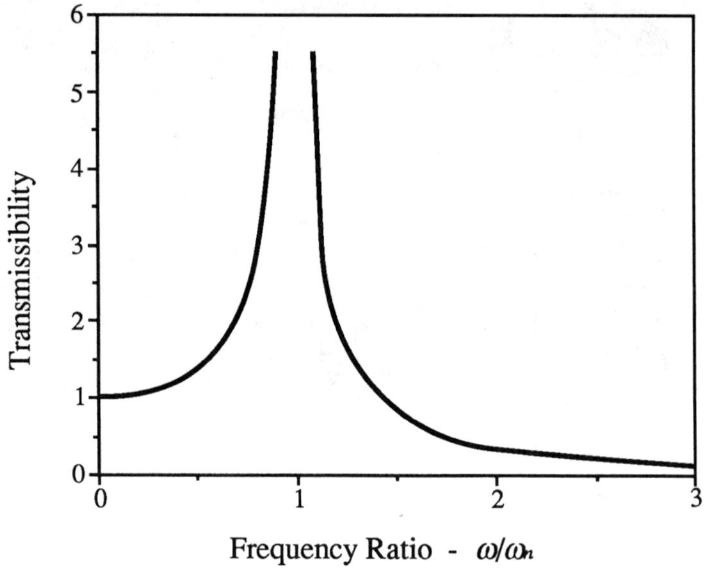


**Figure 2.7** Transmissibility as a Function of Frequency Ratio for a Single Degree-of-Freedom, Undamped, Forced Vibration System

Again considering only the steady-state solution, the motion transmissibility, $T_M$, is given by

$$T_M = \frac{x}{u_o \sin \omega t} = \left( 1 - \frac{\omega^2}{\omega_n^2} \right)^{-1} \tag{2.19}$$

where $T_M$ is unbounded for a frequency ratio of unity. Note that for this model, the force and motion transmissibilities are identical. Thus, the plot in Figure 2.7 applies to motion transmissibility, as well as to force transmissibility.

## Damped Forced Vibration Model

A viscous damping element is now added to the models of Figure 2.6, resulting in a single-degree-of-freedom forced vibration system with viscous damping, as illustrated by the models of Figure 2.8. The forcing function, as before, can be represented as either a force applied to the mass (Figure 2.8a) or a prescribed motion of the foundation (Figure 2.8b).

For the case in which a sinusoidal forcing function is applied to the mass, the governing equation becomes

$$m\ddot{x} + b\dot{x} + kx = F_o \sin \omega t \tag{2.20}$$

where $F_o \sin\omega t$ is the applied force. For reasons given above, only the steady-state solution of Equation (2.20) is considered. In general, $b$ is not negative, although it may be zero. For positive values of $b$, there exists a phase angle between the forcing function $F_o \sin\omega t$ and the steady-state response $x$. This response is given by

$$x = R \sin (\omega t - \theta) \tag{2.21}$$

where

$$R = \frac{F_o}{k}\left[\left(1 - \frac{\omega^2}{\omega_n^2}\right)^2 + (2\zeta\omega/\omega_n)^2\right]^{-1/2} \tag{2.22}$$

and

$$\theta = \text{Arctan}\left[(2\zeta\omega/\omega_n)\left(1 - \frac{\omega^2}{\omega_n^2}\right)^{-1}\right] \tag{2.23}$$

and $\zeta$ and $\omega_n$ are as defined previously.

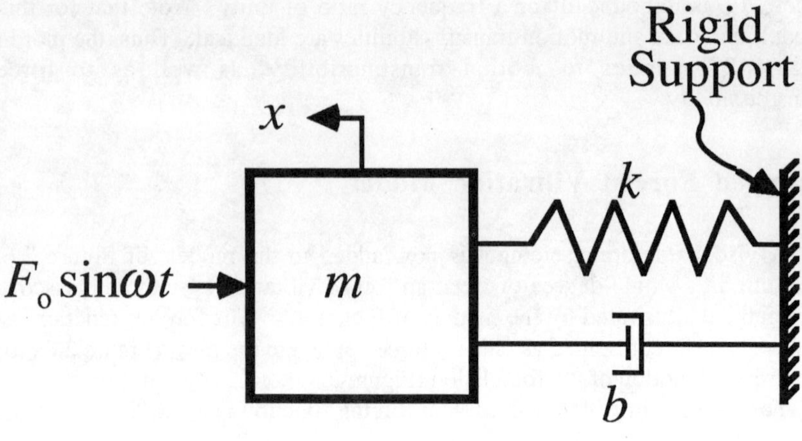


a) with force applied to mass

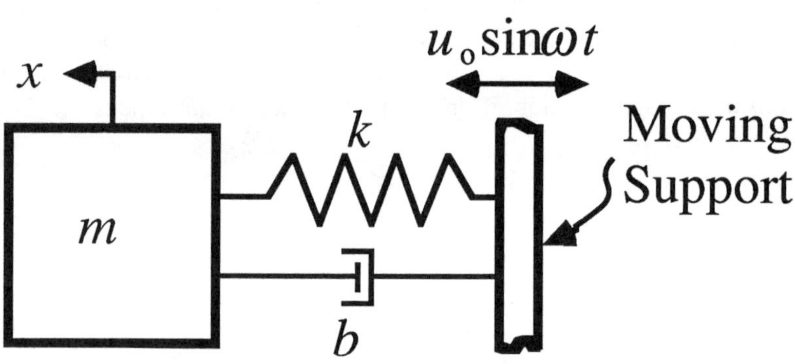


b) with prescribed motion of foundation

**Figure 2.8** Single Degree-of-Freedom Forced Vibration Model with Viscous Damping

The relationship between the force transmitted to the foundation and the applied force is more complex than for the undamped system presented above. For the damped system, both the spring and damper transmit force to the foundation. The total force transmitted to the foundation $F_T$ is given by

$$F_T = b\dot{x} + kx \tag{2.24}$$

where the two terms on the right are ninety degrees out of phase. The ratio of transmitted force, $F_T$, to the magnitude of the applied force, $F_o$, becomes

$$\frac{F_T}{F_o} = T \sin(\omega t - \psi) \tag{2.25}$$

where

$$T = \sqrt{\frac{1 + (2\zeta\omega/\omega_n)^2}{(1 - \omega^2/\omega_n^2)^2 + (2\zeta\omega/\omega_n)^2}} \tag{2.26}$$

$$\psi = \text{Arctan}\left[\frac{2\zeta(\omega/\omega_n)^3}{1 - (\omega^2/\omega_n^2) + (2\zeta\omega/\omega_n)^2}\right] \tag{2.27}$$

and $T$ and $\psi$ are referred to as the transmissibility and the phase angle, respectively. Note that for positive values of $\zeta$, $T$ is bounded for all values of $\omega$. Thus, the existence of damping is necessary to control vibration at a system resonance, as can be seen by comparing Equation (2.26) to Equation (2.16). Plots of normalized response, $R/(F_o/k)$, and phase angle, $\theta$, are presented as functions of the frequency ratio for several values of $\zeta$ in Figures 2.9 and 2.10, respectively.

The maximum value of the displacement, $x$, from Equation (2.21), occurs for a forcing frequency equal to $[\omega_n(1 - 2\zeta^2)^{1/2}]$. Note that for real values of $\zeta$, the maximum steady-state response, in terms of displacement, always occurs for forcing frequencies less than the undamped natural frequency. This is generally true for uni-directional vibratory systems. As will be seen in the next section, this is not true for lateral vibration of rotating machines.

For the case in which the foundation has a prescribed sinusoidal motion (Figure 2.8b), the governing equation becomes

$$m\ddot{x} + b\dot{x} + kx = bu_o\omega \cos \omega t + ku_o \sin \omega t \tag{2.28}$$

where $u_o \sin\omega t$ is the prescribed displacement of the foundation. Consequently, $u_o\omega \cos\omega t$ is the related prescribed velocity of the foundation, since the velocity is the time derivative of the displacement. Again, considering only the steady-state solution to Equation (2.28), the response is given by

$$x = Tu_o \sin(\omega t - \psi) \tag{2.29}$$

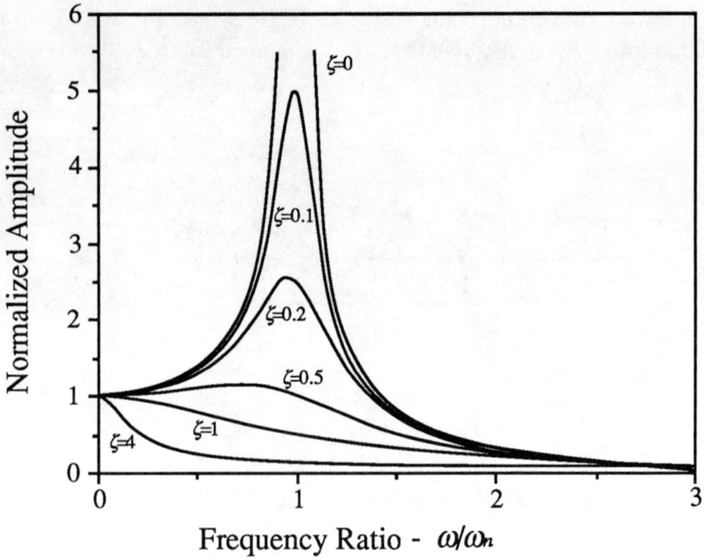


**Figure 2.9** Normalized Response of a Single Degree-of-Freedom Forced Vibration System with Viscous Damping, as a Function of the Frequency Ratio for Several Values of the Damping Ratio

where $T$ and $\psi$ are defined by Equations (2.26) and (2.27), respectively. Analogous to Equation (2.25), the motion transmissibility and phase angle are defined from

$$\frac{x}{u_o} = T \sin\left(\omega t - \psi\right) \tag{2.30}$$

using Equation (2.29) for $x$. Thus, the force transmissibility and phase angle are identical to the motion transmissibility and phase angle for the corresponding forms of the forced vibration.

In general, practical vibratory systems have more than one degree of freedom. Practical systems can usually be represented with reasonable accuracy by a number of spring, mass and damper elements, connected together in some fashion, if a sufficient number of elements are used. This is often referred to as a lumped parameter representation. Real systems may exhibit free or forced vibration, or a combination of both. Synchronous lateral vibration of rotating machines, which is of primary interest in this book, is a form of forced vibration where centrifugal forces, due to mass unbalance, provide the forcing function. This is discussed in detail below. A more complete discussion of multiple degree-of-freedom systems is beyond the scope of this book.

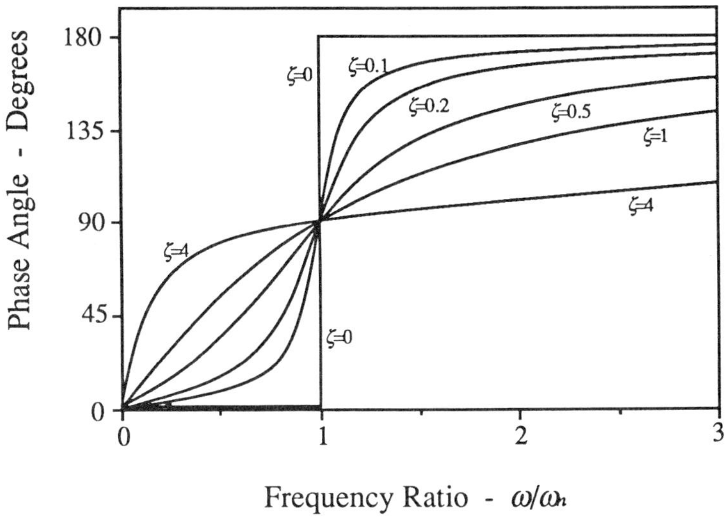


**Figure 2.10** Phase Angle of a Single Degree-of-Freedom Forced Vibration System with Viscous Damping, as a Function of the Frequency Ratio for Several Values of the Damping Ratio

## Basic Theory of Lateral Rotor Vibrations

As indicated in the first chapter, rotor vibrations may be in the form of lateral, axial, or torsional vibrations (see Figure 1.1). These vibrations may also be either synchronous or nonsynchronous. As rotor mass unbalance will result in only synchronous lateral vibration, this is the only form of rotor vibration that is of interest in this book. Consequently, the basic rotordynamic theory presented in this section concerns lateral vibration only, with emphasis on the synchronous frequency components.

Early in the twentieth century, H.H. Jeffcott [68] presented an analysis of a simple, concentrated-mass, flexible rotor including the effect of external damping. Although the model is very simple, this analysis demonstrates the effect of the phase shift of the rotor's center of mass as its critical speed is traversed. The analysis was proved to be valid and has subsequently formed the basis of later, more detailed, analyses of synchronous vibration in damped rotors. In this section, Jeffcott's analysis is briefly reviewed and similar analyses are performed for simple conical and combined conical-translational whirl. The effect of flexible, anisotropic supports is also briefly discussed.

## Translational Jeffcott Analysis

Jeffcott's original analysis is based on the rotor model presented in Figure 2.11. This model consists of a longitudinally and laterally symmetric rotor, with a single concentrated mass in the center, supported by a flexible, massless shaft which is in turn mounted on simple supports. This analysis can be applied to the rotor model presented in Figure 2.12, with no modifications whatsoever, so long as the values of the model parameters (i.e., mass, stiffness, damping, and eccentricity) are the same for both models.

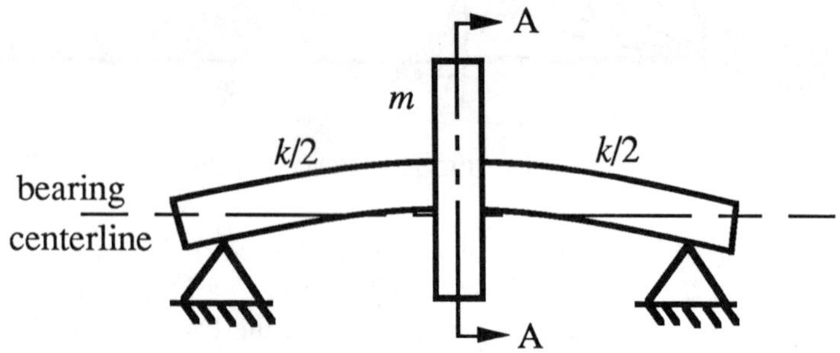


**Figure 2.11** Rotor Model Used for Original Jeffcott Analysis

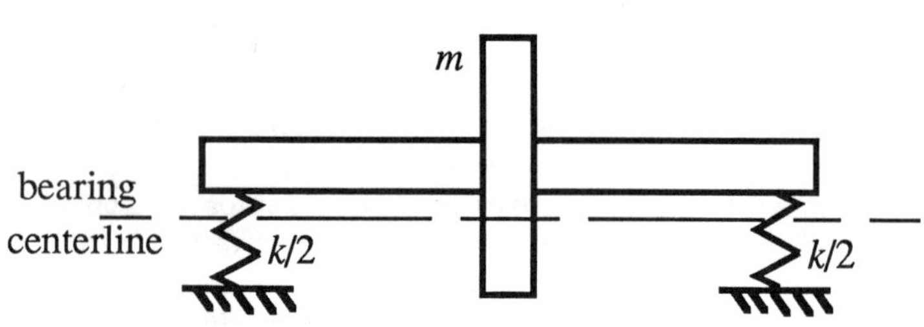


**Figure 2.12** Analogous Rotor Model for Original Jeffcott Analysis

An end view of the rotor model is presented in Figure 2.13, where the origin of the stationary (i.e., non-rotating) coordinate system $XY$ is located at the undeflected centerline of the rotor. For a vertical rotor, the undeflected centerline is coincident with the line of bearing centers. For a horizontal rotor, these two lines differ by the static gravity sag of the rotor. The point $E$ represents the deflected centerline of the rotor, while the point $M$ represents the location of the center of mass of the rotor. Since the coordinate system is not rotating, the sketch in Figure 2-13 represents an instant of time when the rotor is at an angle $\omega t$ from the $X$ axis. Rotation of the rotor is counter-clockwise. The coordinates of the point $E$ at that instant in time are $x$ and $y$. The coordinates of the point $M$ at the same instant in time are $x'$ and $y'$, related to $x$ and $y$ by

$$x' = x + a \cos \omega t \tag{2.31}$$

$$y' = y + a \sin \omega t \tag{2.32}$$

where $a$ is the fixed distance between points $E$ and $M$. This fixed distance is known as the mass eccentricity.

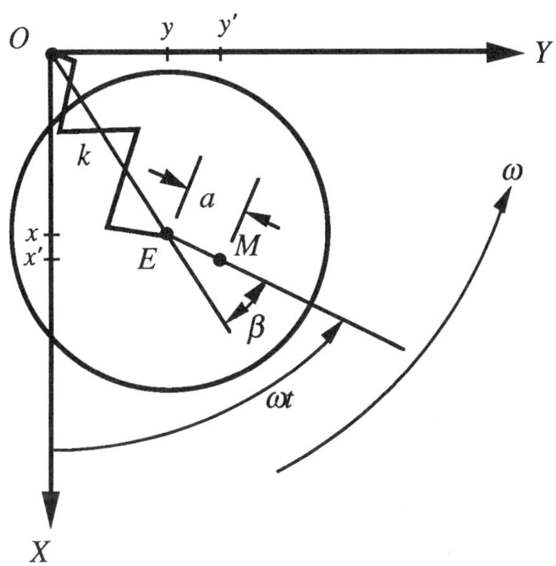


**Figure 2.13** Cross Sectional View of Jeffcott Rotor Model (Section A-A)

The differential equations of motion for the rotor are given by

$$m\ddot{x}' + b\dot{x} + kx = 0 \tag{2.33}$$

$$m\ddot{y}' + b\dot{y} + ky = 0 \tag{2.34}$$

where $m$ is the mass of the rotor and $b$ and $k$ are the external linear viscous damping and stiffness, respectively. The dots again represent derivatives taken with respect to time. In the $X$-direction, the D'Alembert force, $m\ddot{x}'$, is proportional to the acceleration in the $X$-direction of the point $M$. However, the damping and stiffness act at the point $E$ and, therefore, result in forces proportional to $\dot{x}$ and $x$, respectively. The forces in the $Y$-direction act in a similar fashion. Substituting Equations (2.31) and (2.32) into Equations (2.33) and (2.34), to eliminate $x'$ and $y'$, and rearranging terms gives Equations (2.35) and (2.36), assuming $\omega$ is not a function of time.

$$m\ddot{x} + b\dot{x} + kx = ma\omega^2 \cos \omega t \tag{2.35}$$

$$m\ddot{y} + b\dot{y} + ky = ma\omega^2 \sin \omega t \tag{2.36}$$

Note that at this point the rotor system is assumed to be symmetric with respect to the $XY$ coordinate system. That is, the parameters of damping and stiffness are assumed to have equivalent values in the $X$ and $Y$ directions. The effect of anisotropic supports is considered later.

There are various ways to solve Equations (2.35) and (2.36), including the use of Laplace transforms. However, the simplest method of solution is to plug an assumed solution into the differential equations and solve for the constants, since the form of the solution is obvious from Figure 2.13. Jeffcott solves these equations for both the transient (general) and steady-state (particular) solutions. However, since no sources of instability are present in this model, assuming positive damping, the transient (nonsynchronous) solutions will necessarily decay to zero. Therefore, only the steady-state (synchronous) solutions are of interest here. From Figure 2.13, it is clear that

$$x = A \cos (\omega t - \beta) \tag{2.37}$$

$$y = A \sin (\omega t - \beta) \tag{2.38}$$

where $A$ is the distance from $0$ to $E$ and $\beta$ is known as the phase angle. Substituting Equations (2.37) and (2.38) and their derivatives into Equations (2.35) and (2.36) and solving for the constants $A$ and $\beta$ gives

$$A = \frac{ma\omega^2}{\sqrt{\left(k - m\omega^2\right)^2 + b^2\omega^2}} = \frac{a\left(\omega/\omega_n\right)^2}{\sqrt{\left(1 - \omega^2/\omega_n^2\right)^2 + 4\left(\zeta\omega/\omega_n\right)^2}} \tag{2.39}$$

$$\beta = \arctan \frac{b\omega}{k - m\omega^2} = \arctan\left(\frac{2\zeta\omega/\omega_n}{1 - \omega^2/\omega_n^2}\right) \quad \text{where} \quad 0 \le \beta \le \pi \tag{2.40}$$

where $\zeta$ and $\omega_n$ are the same as for uni-directional vibration. Plots of normalized response $(A/a)$ are presented as functions of frequency ratio for several values of $\zeta$ in Figure 2.14. Note that Equation (2.40) is identical to Equation (2.23) and thus plots of $\beta$ will be identical to those of $\theta$ in Figure 2.10.

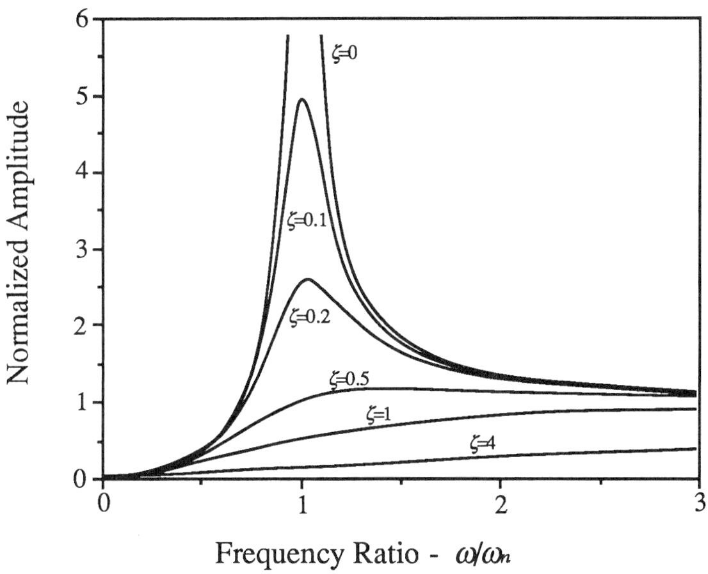


**Figure 2.14** Normalized Amplitude of Jeffcott Rotor Model as a Function of Frequency Ratio for Several Values of Damping Ratio

The critical speed of the rotor is defined as the speed at which the value of $A$ is a maximum. The equation for the critical speed, $\omega_{cr}$, is derived by differentiating the equation for $A$ with respect to $\omega$, setting it equal to zero, and solving for $\omega$, resulting in

$$\omega_{cr} = \frac{2k}{\sqrt{4mk - 2b^2}} = \frac{\omega_n}{\sqrt{1 - 2\zeta^2}} \tag{2.41}$$

It is apparent from Equation (2.41) and Figure 2.14 that the effect of increasing the damping is to increase the critical speed.

## Conical Jeffcott Analysis

A similar analysis can be conducted for a rotor subjected to a conical mode of vibration. This analysis is performed for the rotor model illustrated in Figure 2.15. This rotor model is longitudinally symmetric. That is, a centrally located disk is supported by bearings of equal stiffness. The disk has a mass, $m$, and polar and transverse moments of inertia, $I_p$ and $I_t$, respectively. The analysis can also be applied to the rotor model presented in Figure 2.16, with no modifications whatsoever, so long as the values of the model parameters are the same for both models.

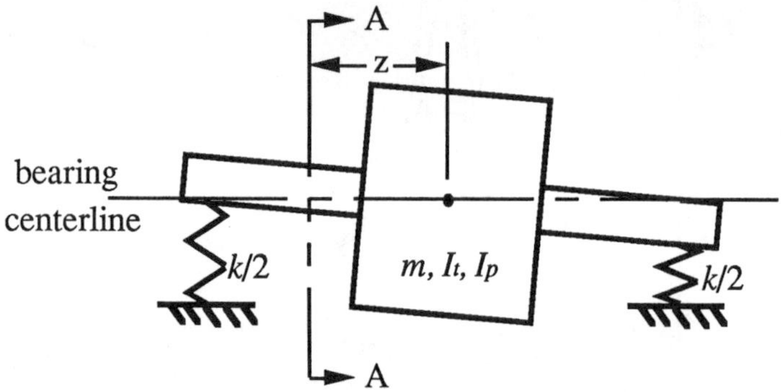


**Figure 2.15** Rotor Model for Conical Jeffcott Analysis

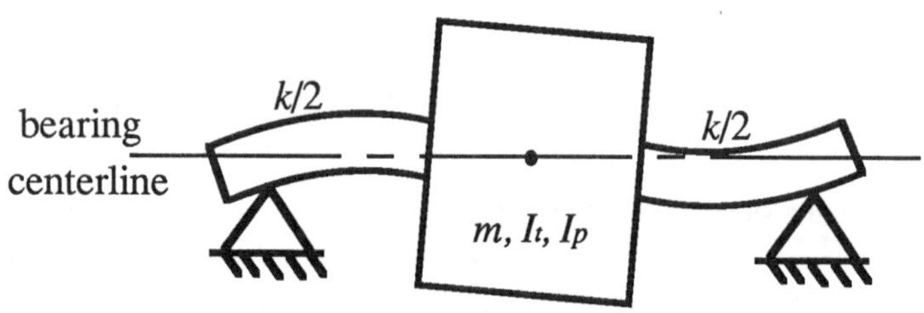


**Figure 2.16** Analogous Rotor Model for Conical Jeffcott Analysis

A view of the rotor in Figure 2.15 at section A-A is presented in Figure 2.17 in the *XY* coordinate system defined above. The mass axis of the rotor is assumed to be displaced angularly with respect to the undeflected shaft axis, with no eccentricity of the center of gravity. Assuming that the mass shift angle, $\theta_a$, is small, the distance from the undeflected shaft axis (the point $E$ in

Figure 2.17) at any cross-section of the rotor is equal to $\theta_a z$, where $z$ is the axial distance from the center of gravity of the rotor to the cross-section. Then $\theta_x$ and $\theta_y$ represent the $X$ and $Y$ components of $\theta$, the angle between the undeflected shaft axis and the bearing axis. Consequently $x$, $y$, $x'$ and $y'$ in Figure 2.17 are equivalent to the expressions

$$x = \theta_x z \tag{2.42}$$

$$y = \theta_y z \tag{2.43}$$

$$x' = \theta_x z + \theta_a z \cos \omega t \tag{2.44}$$

and

$$y' = \theta_y z + \theta_a z \sin \omega t \tag{2.45}$$

The angle between the mass axis of the rotor and the bearing axis, $\theta'$, may then be represented as having $X$ and $Y$ components

$$\theta_x' = \frac{x'}{z} = \theta_x + \theta_a \cos \omega t \tag{2.46}$$

$$\theta_y' = \frac{y'}{z} = \theta_y + \theta_a \sin \omega t \tag{2.47}$$

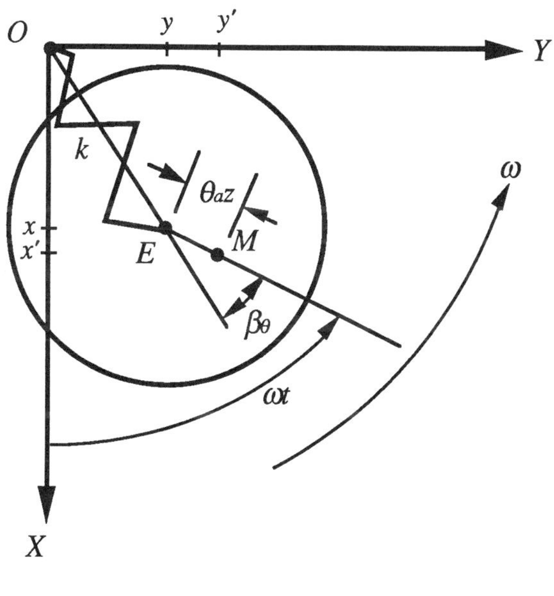


**Figure 2.17** Cross Sectional View of Conical Jeffcott Model (Section A-A)

The differential equations of motion of the rotor are then, including the *gyroscopic effect* [10], represented by Equations (2.48) and (2.49).

$$I_t\ddot{\theta}_x' + I_p\omega\dot{\theta}_y' + b_\theta\dot{\theta}_x + k_\theta\theta_x = 0 \tag{2.48}$$

$$I_t\ddot{\theta}_y' + I_p\omega\dot{\theta}_x' + b_\theta\dot{\theta}_y + k_\theta\theta_y = 0 \tag{2.49}$$

where $b_\theta$ and $k_\theta$ are the angular viscous damping and angular stiffness coefficients, respectively, for the symmetrically supported rotor. Substituting for the primed variables from Equations (2.46) and (2.47), taking the time derivatives and collecting terms results in Equations (2.50) and (2.51).

$$I_t\ddot{\theta}_x + I_p\omega\dot{\theta}_y + b_\theta\dot{\theta}_x + k_\theta\theta_x = (I_t - I_p)\ \theta_a\omega^2 \cos \omega t \tag{2.50}$$

$$I_t\ddot{\theta}_y - I_p\omega\dot{\theta}_x + b_\theta\dot{\theta}_y + k_\theta\theta_y = (I_t - I_p)\ \theta_a\omega^2 \sin \omega t \tag{2.51}$$

Note that these two differential equations are coupled due to the gyroscopic terms. However, it is possible to decouple these equations. Note that $\theta_x$ and $\theta_y$ take the form of Equations (2.52) and (2.53)

$$\theta_x = B \cos (\omega t - \beta_\theta) \tag{2.52}$$

$$\theta_y = B \sin (\omega t - \beta_\theta) \tag{2.53}$$

where $B$ represents the magnitude of angle $\theta$. Their first two time derivatives are simply

$$\dot{\theta}_x = -B\omega \sin (\omega t - \beta_\theta) \tag{2.54}$$

$$\dot{\theta}_y = B\omega \cos (\omega t - \beta_\theta) \tag{2.55}$$

$$\ddot{\theta}_x = -B\omega^2 \cos (\omega t - \beta_\theta) \tag{2.56}$$

$$\ddot{\theta}_y = -B\omega^2 \sin (\omega t - \beta_\theta) \tag{2.57}$$

These equations lead to the equalities

$$\dot{\theta}_y\omega = -\ddot{\theta}_x \tag{2.58}$$

$$\dot{\theta}_x\omega = \ddot{\theta}_y \tag{2.59}$$

Substituting Equations (2.58) and (2.59) into Equations (2.51) and (2.52) results in Equations (2.60) and (2.61), which are completely decoupled from one another.

$$(I_t - I_p)\,\ddot{\theta}_x + b_\theta\dot{\theta}_x + k_\theta\theta_x = (I_t - I_p)\theta_a\omega^2 \cos \omega t \tag{2.60}$$

$$(I_t - I_p)\,\ddot{\theta}_y + b_\theta\dot{\theta}_y + k_\theta\theta_y = (I_t - I_p)\theta_a\omega^2 \sin \omega t \tag{2.61}$$

Equations (2.60) and (2.61) are directly analogous to Equations (2.35) and (2.36) with the substitutions of the angular coefficients and variables. Consequently, the same substitutions may be made in Equations (2.39), (2.40), and (2.41) in order to obtain Equations (2.62) through (2.64), which, together with Equations (2.52) and (2.53) define the conical mode of vibration of the rotor model in Figure 2.15.

$$\begin{aligned} B &= \frac{(I_t - I_p)\theta_a\omega^2}{\sqrt{\left[k_\theta - (I_t - I_p)\omega^2\right]^2 + b_\theta^2\omega^2}} \\ &= \frac{\theta_a(\omega/\omega_{n\theta})^2}{\sqrt{\left(1 - \omega^2/\omega_{n\theta}^2\right)^2 + 4(\zeta_\theta\omega/\omega_{n\theta})^2}} \end{aligned} \tag{2.62}$$

$$\begin{aligned} \beta_\theta &= \arctan \frac{b_\theta\omega}{k_\theta - (I_t - I_p)\omega^2} \\ &= \arctan \frac{2\zeta_\theta\omega/\omega_{n\theta}}{1 - \omega^2/\omega_{n\theta}^2} \qquad \text{where} \qquad 0 \le \beta_\theta \le \pi \end{aligned} \tag{2.63}$$

$$\omega_{cr} = \frac{2k_\theta}{\sqrt{4(I_t - I_p)k_\theta - 2b_\theta^2}} = \frac{\omega_{n\theta}}{\sqrt{1 - 2\zeta_\theta^2}} \tag{2.64}$$

where $\omega_{n\theta}$ and $\zeta_\theta$ are defined the same as $\omega_n$ and $\zeta$, except that $(I_t - I_p)$ is substituted for $m$, $k_\theta$ for $k$, and $b_\theta$ for $b$. Note that both the damping and gyroscopic terms tend to raise the critical speed. This rotor model will also exhibit a separate purely translational critical speed, whose response may be calculated using Jeffcott's original analysis, as described above.

## Combined Jeffcott Analysis

A rotor, similar to that in Figure 2.16 but longitudinally nonsymmetric, will not exhibit purely translatory and conical modes of vibration. Instead, this rotor will have two rigid-body critical speeds whose mode shapes are each a combination of translatory and conical whirl. Generally speaking, the translatory whirl predominates for one of these mode shapes while the conical whirl predominates for the other. Such a rotor model is presented in Figure 2.18. The governing differential equations of motion for this model are presented in Equations (2.65) through (2.68).

$$m\ddot{x}' + b\dot{x} + kx + (b_2l_2 - b_1l_1)\dot{\theta}_x + (k_2l_2 - k_1l_1)\,\theta_x = 0 \tag{2.65}$$

$$I_t\ddot{\theta}_x' + I_p\omega\dot{\theta}_y' + b_\theta\,\dot{\theta}_x + k_\theta\theta_x + (b_2l_2 - b_1l_1)\dot{x} + (k_2l_2 - k_1l_1)x = 0 \tag{2.66}$$

$$m\ddot{y}' + b\dot{y} + ky + (b_2l_2 - b_1l_1)\dot{\theta}_y + (k_2l_2 - k_1l_1)\theta_y = 0 \tag{2.67}$$

$$I_t\,\ddot{\theta}_y' - I_p\omega\dot{\theta}_x' + b_\theta\,\dot{\theta}_y + k_\theta\theta_y + (b_2l_2 - b_1l_1)\dot{y} + (k_2l_2 - k_1l_1)y = 0 \tag{2.68}$$

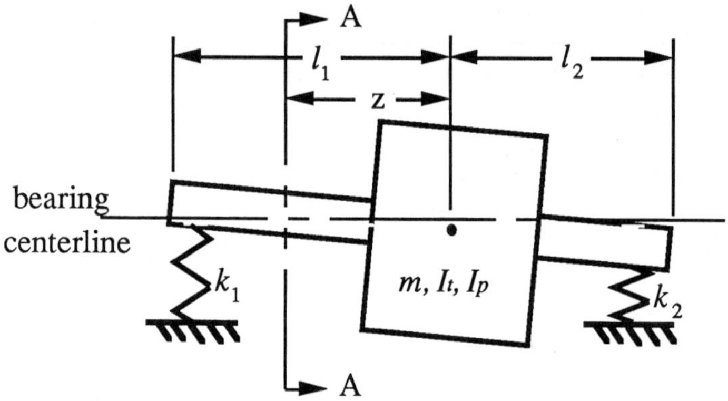


**Figure 2.18** Rotor Model for Coupled Translational-Conical Jeffcott Analysis

These equations are derived using superposition of the translatory and conical unbalance distributions and modes of vibration, where the total resultant vibration is a linear combination of pure translatory and pure conical motion. The non-primed independent variables in equations (2.65) through (2.68) are represented by

$$x = A\cos(\omega t - \beta_t) \tag{2.69}$$

$$y = A\sin(\omega t - \beta_t) \tag{2.70}$$

$$\theta_x = B\cos(\omega t - \beta_c) \tag{2.71}$$

$$\theta_y = B\sin(\omega t - \beta_c) \tag{2.72}$$

where $\beta_t$ and $\beta_c$ represent $\beta$ in Figures 2.13 and 2.17, respectively. The coefficients $A$ and $B$ represent the distance $OE$ and the angle $\theta$, respectively, from the same figures.

Equations (2.65) and (2.66) may be decoupled from Equations (2.67) and (2.68) using the substitutions given by Equations (2.58) and (2.59) as described above. However, Equations (2.65) and (2.67) cannot be decoupled from Equations (2.66) and (2.68) unless the terms in parentheses go to zero. If the conditions in Equations (2.73) and (2.74) are satisfied

$$b_1 l_1 = b_2 l_2 \tag{2.73}$$

$$k_1 l_1 = k_2 l_2 \tag{2.74}$$

then Equations (2.65) and (2.66) decouple and reduce to Equations (2.33) and (2.48), indicating the presence of separate pure translatory and conical modes of vibration. Equations (2.67) and (2.68) are similarly affected. If the conditions in either or both of Equations (2.73) and (2.74) are not satisfied, then Equations (2.65) and (2.67) do not decouple from Equations (2.66) and (2.68), respectively. In either case, Equations (2.65) and (2.66) are decoupled from and analogous to Equations (2.67) and (2.68) and solving the first two equations implies a solution to the latter two equations. Therefore, only Equations (2.65) and (2.66) are solved, from which the solutions of Equations (2.67) and (2.68) are inferred.

Substitutions for $x'$ and $\theta_x'$ from Equations (2.31) and (2.46) and the substitutions from Equations (2.58) and (2.59) are introduced into Equations (2.65) and (2.66) and the terms are rearranged resulting in Equations (2.75) and (2.76).

$$m\ddot{x} + b\dot{x} + kx + \left(b_2 l_2 - b_1 l_1\right)\dot{\theta}_x + \left(k_2 l_2 - k_1 l_1\right)\theta_x = ma\omega^2 \cos \omega t \tag{2.75}$$

$$\left(I_t - I_p\right)\ddot{\theta}_x + b_\theta \theta_x + k_\theta \theta_x + \left(b_2 l_2 - b_1 l_1\right)\dot{x} + \left(k_2 l_2 - k_1 l_1\right)x = \left(I_t - I_p\right)\theta_a \omega^2 \cos \omega t \tag{2.76}$$

The substitutions in Equations (2.77) through (2.80) are implied in Equations (2.75) and (2.76), based on the rotor model in Figure 2.18.

$$b = b_1 + b_2 \tag{2.77}$$

$$k = k_1 + k_2 \tag{2.78}$$

$$b_\theta = b_1 l_1^2 + b_2 l_2^2 \tag{2.79}$$

$$k_\theta = k_1 l_1^2 + k_2 l_2^2 \tag{2.80}$$

In addition, the substitutions in Equations (2.81) through (2.83) are made in order to reduce Equations (2.75) and (2.76) to Equations (2.84) and (2.85).

$$I = I_t - I_p \tag{2.81}$$

$$b_{21} = (b_2 l_2 - b_1 l_1)/l \tag{2.82}$$

$$k_{21} = (k_2 l_2 - k_1 l_1)/l \tag{2.83}$$

$$m\ddot{x} + b\dot{x} + kx + b_{21} l\dot{\theta}_x + k_{21} l\theta_x = ma\omega^2 \cos \omega t \tag{2.84}$$

$$I\ddot{\theta} + b_\theta \dot{\theta}_x + k_\theta \theta_x + b_{21} l\dot{x} + k_{21} lx = I\theta_a \omega^2 \cos \omega t \tag{2.85}$$

Direct substitution of Equations (2.69) and (2.71) into Equations (2.84) and (2.85) results in four simultaneous transcendental equations for which there is no apparent closed-form solution. However, Equations (2.69) and (2.71) can be rewritten as

$$x = G \sin \omega t + D \cos \omega t \tag{2.86}$$

$$\theta_x = E \sin \omega t + F \cos \omega t \tag{2.87}$$

where

$$G = A \sin \beta_t \tag{2.88}$$

$$D = A \cos \beta_t \tag{2.89}$$

$$E = B \sin \beta_c \tag{2.90}$$

$$F = B \cos \beta_c \tag{2.91}$$

Substituting Equations (2.86) and (2.87) into Equations (2.84) and (2.85) results in Equations (2.92) through (2.95), a set of four simultaneous linear equations in $G, D, E$, and $F$.

$$\left(k - m\omega^2\right)G - b\omega D + k_{21}lE - b_{21}l\omega F = 0 \tag{2.92}$$

$$b\omega G + \left(k - m\omega^2\right)D + b_{21}l\omega E + k_{21}lF = ma\omega^2 \tag{2.93}$$

$$k_{21}lG - b_{21}l\omega D + \left(k_\theta - I\omega^2\right)E - b_\theta\omega F = 0 \tag{2.94}$$

$$b_{21}lG + k_{21}lD + b_\theta\omega E + \left(k_\theta - I\omega^2\right)F = I\theta_a\omega^2 \tag{2.95}$$

The expressions for the coefficients $G, D, E$, and $F$ which satisfy Equations (2.92) through (2.95) are presented in Equations (2.96) through (2.99),

$$G = \left\{ ma\omega^2\left[b_\theta\omega\left(S_1 + b_\theta b\omega^2\right) + k_{\theta I}S_\theta\right] + I\theta_a\omega^2\left[b_{21}l\omega\left(T_1 + T_2\right) - k_{21}lT_3\right]\right\}/H \tag{2.96}$$

$$D = \left\{ ma\omega^2\left[b_\theta\omega S_x + k_{\theta I}\left(k_{xm}k_{\theta I} - S_1\right)\right] + I\theta_a\omega^2\left[-b_{21}l\omega T_3 + k_{21}l\left(T_1 - T_2\right)\right]\right\}/H \tag{2.97}$$

$$E = \left\{ ma\omega^2\left[b_{21}l\omega\left(T_1 + T_2\right) - k_{21}lT_3\right] + I\theta_a\omega^2\left[b\omega\left(S_1 + b_\theta b\omega^2\right) + k_{xm}S_x\right]\right\}/H \tag{2.98}$$

$$F = \left\{ ma\omega^2\left[-b_{21}l\omega T_3 + k_{21}l\left(T_1 - T_2\right)\right] + I\theta_a\omega^2\left[b\omega S_\theta + k_{xm}\left(k_{xm}k_{\theta I} - S_1\right)\right]\right\}/H \tag{2.99}$$

where

$$S_1 = k_{21}^2 l^2 - b_{21}^2 l^2 \omega^2 \tag{2.100}$$

$$S_\theta = k_{\theta I}b\omega - 2k_{21}b_{21}l^2\omega \tag{2.101}$$

$$S_x = k_{xm} b_\theta \omega - 2k_{21} b_{21} l^2 \omega \tag{2.102}$$

$$T_1 = k_{21}^2 l^2 + b_{21}^2 l^2 \omega^2 \tag{2.103}$$

$$T_2 = k_{xm} k_{\theta I} - b b_\theta \omega^2 \tag{2.104}$$

$$T_3 = k_{xm} b_\theta \omega + k_{\theta I} b \omega \tag{2.105}$$

$$H = \left(k_{xm}^2 + b^2 \omega^2\right)\left(k_{\theta I}^2 + b_\theta^2 \omega^2\right) + T_1^2 - 2T_2 S_1 - 4T_3 b_{21} k_{21} l^2 \omega \tag{2.106}$$

$$k_{xm} = k - m\omega^2 \tag{2.107}$$

$$k_{\theta I} = k_\theta - I\omega^2 \tag{2.108}$$

It is then possible to derive $x$ and $\theta_x$ in the form of Equations (2.69) and (2.71) by using the relations

$$A = \sqrt{G^2 + D^2} \tag{2.109}$$

$$\beta_t = \arctan\,(G/D) \tag{2.110}$$

$$B = \sqrt{E^2 + F^2} \tag{2.111}$$

$$\beta_c = \arctan\,(E/F) \tag{2.112}$$

where

$$0 \le \beta_t \le \pi \quad \text{and} \quad 0 \le \beta_c \le \pi$$

The resulting expressions for $A$ and $B$ may then be differentiated with respect to $\omega$, in order to calculate the critical speeds of the rotor, as described above. The algebraic manipulation required to apply Equations (2.109) through (2.112) and calculate the critical speeds is quite involved. For rotors where the damping is expected to be very small the error in the predicted values of critical speed introduced by neglecting damping is also very small. In light of the fact that the values of rotor mass, inertia and support stiffness used for these calculations are often only approximate, the error mentioned above becomes even less significant. Neglecting damping simplifies Equations (2.96) through (2.106) tremendously, and Equations (2.109) through (2.112) become

$$A = \left[ma\omega^2 k_{\theta I} - I\theta_a\omega^2 k_{21}l\right] / \left[k_{xm}k_{\theta I} - k_{21}^2 l^2\right] \tag{2.113}$$

$$\beta_t = 0 \text{ or } \pi \text{ (i.e., below or above } \omega_{cr}) \tag{2.114}$$

$$B = \left[I\theta_a\,\omega^2 k_{xm} - ma\omega^2 k_{21}l\right] / \left[k_{xm}k_{\theta I} - k_{21}^2 l^2\right] \tag{2.115}$$

$$\beta_c = 0 \text{ or } \pi \text{ (below or above } \omega_{cr}) \tag{2.116}$$

For an undamped rotor, at the critical speeds the amplitude of vibration becomes infinite. This occurs when the denominators of Equations (2.113) and (2.115) are equal to zero, or when

$$k_{xm}k_{\theta I} - k_{21}^2 l^2 = 0 \tag{2.117}$$

Obviously, if $k_{21}$ is equal to zero, a critical speed occurs where $k_{xm}$ or $k_{\theta I}$ is equal to zero, which represents the decoupled mode case discussed above. Otherwise, solving Equation (2.117) for $\omega_{cr}^2$,

$$\omega_{cr}^2 = \frac{k}{2m} + \frac{k_\theta}{2I} \pm \sqrt{\left(\frac{k}{2m} - \frac{k_\theta}{2I}\right)^2 + \frac{k_{21}^2 l^2}{Im}} \tag{2.118}$$

where only the positive square roots of $\omega_{cr}^2$ are of interest. Equation (2.118) may be rewritten as Equation (2.119) from which it becomes clear that $\omega_{cr}^2$ must be real and positive.

$$\omega_{cr}^2 = \frac{k}{2m} + \frac{k_\theta}{2I} \pm \sqrt{\left(\frac{k}{2m} - \frac{k_\theta}{2I}\right)^2 - \frac{k_1 k_2 l^2}{mI}} \tag{2.119}$$

Regardless of the complexity of its geometric configuration, a rigid rotor can be represented exactly by a mass, transverse and polar moments of inertia and a center of gravity location. Thus, the mass unbalance of a rigid rotor can also be represented exactly by two parameters, $a$ and $\theta_a$. Since $a$ can be reduced to zero with a single correction mass located at an arbitrary axial location and $\theta_a$ can be reduced to zero, with an appropriate pair of correction masses at appropriate axial locations, it seems reasonable to assume that any arbitrary unbalance in a rigid rotor can be corrected with a maximum of three correction masses. In fact, since the correction mass for $a$ can be located in the same diametral plane as one of the correction masses for $\theta_a$, and the two correction masses vectorially summed to form a single correction mass, no more than two correction masses are actually required to balance a rigid rotor.

A flexible rotor, however, generally has more than two degrees of freedom and often requires more than two correction masses. In essence, when a rotor bends, the mass centerline also bends and the mass unbalance can no longer be represented by $a$ and $\theta_a$. Thus, the rigid rotor corrections which have been made to reduce $a$ and $\theta_a$ to zero, do not cancel the new mass unbalance distribution that occurs when the rotor bends.

In the linear analysis of rotating machines, two assumptions are made which are central to rotor balancing: (1) Synchronous rotor response is linearly related to mass unbalance, and (2) the effect of individual unbalances may be superposed. These assumptions, while not strictly adhered to in any real nonlinear system, are generally very accurate for relatively small changes in unbalance.

## Synchronous Rotor Response Notation

A complex notation is often used to present the in-phase and out-of-phase components of rotordynamic response. The use of this complex notation simplifies the modeling and analytical procedures for realistic rotor systems and is convenient when considering rotor mass unbalance. Most rotor balancing procedures employ some form of complex notation.

Rotor response may be represented using either a stationary or rotating frame of reference, as illustrated in Figure 2.19. The stationary coordinate system, $XY$, is fixed in space, while the rotating coordinate system, $UV$, is fixed to the bearing centerline and rotates with the rotor. For a constant speed of rotation and synchronous vibration only, which is the condition of interest for rotor balancing, the location of the rotor axis in the rotating coordinate system, represented by the directional vector (i.e., as distinguished from a single dimension matrix) $\hat{R}$ in Figure 2.19, does not change with time if the rotor supports are symmetric. This is referred to as a circular orbit. This directional vector may be specified by either the orthogonal components $u$ and $v$, or the magnitude $r$ and angle $\phi$, where

$$r = \sqrt{u^2 + v^2} \tag{2.120}$$

and

$$\phi = \arg\,(u + iv) \tag{2.121}$$

The superior caret is used to indicate complex variables and the two-dimensional, directional vector $\hat{R}$ may be represented in complex notation by

$$\hat{R} = u + iv = r\,e^{i\phi} \tag{2.122}$$

where

$$e^{i\phi} = \cos\phi + i\,\sin\phi \tag{2.123}$$

$$i = \sqrt{-1} \tag{2.124}$$

In standard nomenclature, $u$ and $v$ are referred to as the real and imaginary components of $\hat{R}$, and $r$ and $\phi$ are referred to as the magnitude and phase angle of $\hat{R}$, respectively. In essence, $u$ and $v$ may be thought of as cartesian coordinates while $r$ and $\phi$ are thought of as polar coordinates.

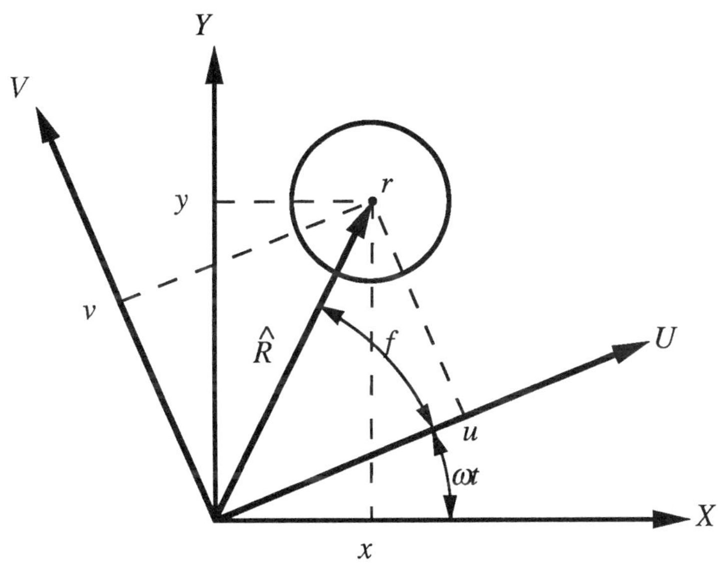


**Figure 2.19** Stationary and Rotating Coordinate Systems for Measurement of Synchronous Rotor Vibration

The directional vector $\hat{R}$ is represented in the stationary coordinate system in a similar form as Equation (2.122) by

$$\hat{R} = r\, e^{i\gamma} \tag{2.125}$$

where

$$\gamma = \omega t + \phi \tag{2.126}$$

for a leading phase convention, as discussed below. Equation (2.125) may also be written in the form

$$\hat{R} = r\, e^{i\phi}\, e^{i\omega t} = (u + iv)\, e^{i\omega t} \tag{2.127}$$

Equation (2.127) may be further expanded into either of the following forms,

$$\hat{R} = (u \cos \omega t - v \sin \omega t) + i(u \sin \omega t + v \cos \omega t) \tag{2.128}$$

or

$$\hat{R} = r \cos (\omega t + \phi) + ir \sin (\omega t + \phi) \tag{2.129}$$

The real and imaginary components of Equations (2.128) and (2.129) represent the time dependent values of $x$ and $y$ of Figure 2.19, respectively.

In general, actual rotor vibration is measured using a vibration sensor which is fixed in space and is, by common convention, located in the $X$-direction. Thus, the measured vibration, $x$, can be represented by the real part of Equation (2.128) or (2.129).

$$x = u \cos \omega t - v \sin \omega t = r \cos (\omega t + \phi) \tag{2.130}$$

The real part of Equation (2.129) is identical in form to Equation (2.37) from the rotordynamic analysis, except for the sign of the phase angle, $\phi$ (or $\beta$). This difference in sign occurs because $\beta$ is taken to be positive in a direction opposite to rotation (referred to as a lagging phase angle) while $\phi$ is taken to be positive in the same direction as rotation (referred to as a leading phase angle). For the remainder of this volume, leading phase angles are assumed. However, so long as the phase angles and the balance mass angles are taken to be positive in the same direction, the balancing calculations for most balancing methods are identical for leading and lagging phase angles.

The usual convention is to write the synchronous rotor response $x$, as

$$\hat{x} = x_c + ix_s = r\, e^{i\phi} \tag{2.31}$$

where $x_c$ and $x_s$ are used to represent $u$ and $v$, respectively, and Equation (2.131) is actually a shorthand form for

$$x = \text{Real}\left[\left(x_c + ix_s\right) e^{i\omega t}\right] = \text{Real}\left[r\, e^{i\phi} e^{i\omega t}\right] \tag{2.132}$$

Thus, synchronous rotor vibration is generally represented using the complex notation of Equation (2.131) in either cartesian or polar form. This convention is followed for the remainder of this book. The locations and sizes of discrete mass unbalances are identified by

$$\hat{u} = u_c + iu_s = r_u\, e^{i\phi_u} \tag{2.133}$$

where $u_c$ and $u_s$ are the components in the $U$ and $V$ directions and $r_u$ and $\phi_u$ are the magnitude and phase angle relative to the $UV$ coordinate system, respectively. The form of Equation (2.133) is exactly analogous to Equation (2.131).

If another vibration sensor is located in the $Y$-direction of Figure 2.19, the vibration which is measured is represented by the imaginary part of Equation (2.128) or (2.129) for the circular orbit case (symmetric rotor supports). In the notation of Equation (2.131), $\hat{y}$ is given by

$$\hat{y} = y_c + iy_s = r_y\, e^{i\phi_y} \tag{2.134}$$

where for a circular orbit

$$y_c = x_s \tag{2.135}$$

$$y_s = -x_c \tag{2.136}$$

$$r_y = r \tag{2.137}$$

$$\phi_y = \phi + \pi/2 \tag{2.138}$$

and the phase angles are measured in radians.

## Elliptical Orbit Response

For the more complicated situation of anisotropic bearing supports, which results in elliptical orbits, Equations (2.131) and (2.134) remain valid, but the relations in Equations (2.135) through (2.138) no longer hold. Thus, the notation and calculations required for balancing a rotor with an elliptical orbit is not different than that for a circular orbit, when using $x$ and $y$ vibration measurements individually. However, treating the $x$ and $y$ vibration measurements as independent data is not entirely accurate and can lead to difficulties when attempting to reduce the two simultaneously in a single balancing run. It is preferable to combine the two vibration measurements into a single independent data sample. Lund [85] has shown that an elliptical orbit of this type is due to a combination of forward and backward precessing synchronous whirl. Forward precession indicates that the rotor whirl is occurring in the same direction as the shaft rotation. Conversely, backward precession indicates the direction of whirl is opposite that of rotation. Thus, an elliptical orbit is actually formed from two circular orbits precessing in opposite directions. Since rotor unbalance does no work on the backward precession component [85], only the forward precession component is of interest in balancing. Based on a suggestion from Lund [86], a method has been developed for extracting the forward precession component of the synchronous rotor whirl from a pair of vibration readings at a single axial plane. This is done by subtracting from one of the vibration vectors an angle equal to the angle from the corresponding vibration sensor to the other sensor measured in the direction of shaft rotation (for a leading phase angle). The resulting vector is then vector averaged with the other vibration vector, by halving the vector sum, to give the forward precession component of the synchronous vibration. While this may be done for any sensor separation angle, other than 0 or 180 degrees, the best results are usually obtained with sensor separation angles in the neighborhood of 90 degrees. This means of extracting the forward precession component can be used with virtually any rotor balancing procedure, so long as data is available from appropriately located vibration sensors.

# Chapter 3
# Review of Literature on Rotor Balancing

While the necessity of rotor balancing was demonstrated by Jeffcott in his classic paper [68], published in 1919, the first significant contributions to the rotor balancing literature did not appear until about 1930. Prior to the 1950s, the balancing literature was concerned with the balancing of rigid rotors and, in a few cases, very simple flexible ones. There is no documentation during this period of any systematic balancing procedure using more than two balancing planes. The first flexible rotors of significance to be built were steam turbine rotors. Initially, these rotors were balanced using simple, rigid-rotor procedures. However, as these rotors became more flexible, and were often operated supercritically, the existing balancing procedures became inadequate. This led to the development of a number of balancing methods specifically designed for flexible rotors.

Flexible rotor balancing procedures can generally be divided into two groups, modal balancing and influence coefficient balancing, as defined in Chapter 1, of which modal balancing was developed first. The development of modal balancing began in the early 1950s. As very limited instrumentation and computational tools were available at that time, a balancing method was needed that did not depend heavily on such tools. Modal balancing fit naturally into these requirements as only simple calculations are required and operator insight is the primary ingredient, rather than large quantities, and quality, of vibration data.

Influence coefficient balancing was developed some years later, made possible by improvements in instrumentation and the introduction of the digital computer. Consequently, the use of large quantities of high quality data was substituted for operator insight as the central component in the balancing procedure. Subsequently, the Unified Balancing Approach was developed as an empirical method, in the mold of influence coefficient balancing. It was designed to take advantage of the modal nature of rotor response, so as to avoid some of the difficulties of influence coefficient balancing.

This chapter presents an overview of the rotor balancing literature, with particular emphasis on flexible rotor balancing. Specific reference is made to a number of the more significant contributions. A fairly comprehensive review of the flexible rotor balancing literature prior to 1972 was presented by Rieger [114] and has been an invaluable source in the assembly of this literature review. A discussion of the general and rigid-rotor balancing literature is followed by separate discussions of the literature on modal and influence coefficient balancing and comparisons and combined methods, including the Unified Balancing Approach. Technical discussions of the particular balancing methods are presented in subsequent chapters.

## General and Rigid Rotor Balancing Literature

Most mechanical engineering handbooks and general references include some mention of rotor balancing. In most cases, this mention is strictly limited to rigid rotor balancing. A number of technical papers concerned with general and rigid rotor balancing have also been published.

One of the earliest general references to include a discussion of rotor balancing was that of Den Hartog [38] in 1934. Most of the emphasis in this discussion was on various methods and machines used for rigid rotor balancing at the time of publication. However, a brief discussion of flexible rotor balancing was also included. A very simple single plane flexible rotor balancing procedure was proposed.

A similar discussion was presented by Timoshenko [135], in 1928, in which several rigid rotor balancing machines of that era were described in considerable detail. The need for special balancing procedures for flexible rotors was mentioned but no such procedures of a practical nature were proposed.

More recently, similar discussions were presented in both editions of The Shock and Vibration Handbook [62, 63], in 1961 and 1976. While the two editions presented slightly different discussions, the content was basically the same. The primary emphasis was on an updated review of machines and methods for balancing rigid rotors. A brief discussion of flexible rotor balancing was also presented. However, no particular flexible rotor balancing procedures were discussed in any detail.

Discussions of rotor balancing may also be found in some general mechanical engineering references, such as Mark's Handbook for Mechanical Engineers [1]. Such discussions are generally limited to the need for balancing rigid rotors, and a brief description of rigid rotor balancing machines.

In the early 1940s, Kroon published two papers on rotor balancing [74, 75], which were apparently intended as a design guide. In the first of these papers, Kroon described the theory behind synchronous rotor vibration and the need for balancing of both rigid and flexible rotors. In the second paper, he discussed a number of specific rotor balancing machines and methods. While this discussion was primarily concerned with rigid rotor balancing, a graphical method was described for two plane balancing of flexible rotors. He also presented a brief, practically oriented discussion of field balancing.

A number of other papers concerned with general rotor balancing have been published, including papers by Muster and Flores [95], Jackson [67], and Stadelbauer [123]. Muster and Flores compiled rigid rotor balancing criteria from a variety of sources and compared these criteria with the actual criteria used in American industry at the time (1969). Jackson described a procedure for single plane field balancing of rigid or flexible rotors using an oscilloscope lissajous pattern of the rotor orbit. Stadelbauer presented a review of the history of rigid rotor balancing machines.

Gleizer [54] proposed a method for determining allowable unbalance criteria for flexible rotors, analogous to similar, well-accepted methods for rigid rotors. Using this new method, he developed [55] a probablistic approach to estimating the effectiveness of low-speed (i.e. rigid rotor) balancing of flexible rotors. In an example which he presented, the resulting high-speed response, after low-speed balancing, did not even meet his *good* criteria for a majority of the rotors considered.

Van de Vegte and Lake [141] proposed a procedure for balancing rigid rotors during operation which utilizes actively controllable, eccentric disks. They indicated the potential adaptation of such a mechanism to modal balancing of flexible rotors, but provided no details in this regard. Subsequently, Bishop [13] proposed to use this same balancing head design for balancing flexible rotors. This would be done with a single head located, axially, as far as possible from all mode shape nodes. Then, the head would be readjusted, using a simplified procedure also proposed by Bishop, in the vicinity of each critical speed during each run up and run down of the rotor. Gosiewski [58, 59] also promoted the use of balancing heads for balancing flexible rotors. However, unlike Bishop, he proposed using multiple heads in a procedure which might be described as automatic, and continuous, influence coefficient balancing. He defined the procedure and control algorithm for implementing this approach using on-line computer control. Gosiewski also evaluated several balancing head designs, including that of Van de Vegte and Lake, and concluded that cartesian adjustment approaches, which use laterally adjustable orthogonal masses, are superior to polar adjustment approaches, which use rotating eccentric masses.

## Modal Balancing Literature

Modal balancing was developed considerably earlier than influence coefficient balancing. An early balancing technique, akin to modal balancing, was described by Grobel [61] in 1953. This method was a trial mass procedure which was designed specifically for balancing turbine-generator rotors. Grobel also presented detailed descriptions of several practical problems associated with turbine-generator balancing.

Meldal [91] first developed the orthogonality relations governing modal rotor response, and thus laid the theoretical foundation for modal balancing, in 1954. Based on these relations, he outlined the principles of modal balancing, and described the procedure for balancing a rotor through its first three modes. In his discussion, the measurement of rotor whirl was limited to the bearing locations. However, in general, this is not a necessary limitation.

Beginning in 1957, a series of papers on modal balancing have been published by Bishop and his co-workers, particularly Parkinson. These papers have been concerned with both the theory and application of modal balancing. The flexible rotor balancing method developed in these papers is that most often associated with the term *modal balancing*. In the first of these papers [14], Bishop derived a modal series representation of the synchronous whirl of a lightly damped rotor, based on the Jeffcott model analysis. This analysis was extended by Gladwell and Bishop [53] to obtain natural frequencies and mode shapes of an axi-symmetric shaft of nonuniform cross-section on flexible bearings.

The first paper in this series to consider the balancing of flexible rotors was Bishop and Gladwell [15], in which the *modal balancing* method was first proposed. It was shown analytically that *low speed* balancing is entirely inadequate for a flexible rotor. The effects of shaft bow and gravity sag on unbalance response were also computed analytically and found to be, in general, quite small. A number of papers followed in which the *modal balancing* method was extended or evaluated in light of specific practical problems or conditions which may be encountered in the balancing of flexible rotors. Bishop and Parkinson [17] considered the situation in which the resonant speeds for two, or more, modes are not separated sufficiently to obtain reasonably isolated modes. They discussed the disadvantages of *modal balancing* in such a situation, and proposed a modification, adapted from the Kennedy and Pancu [72] method of resonance testing, for such a case.

Lindley and Bishop [80] considered the practical aspects of applying *modal balancing* to large steam turbine rotors in 1963. Actual experience with balancing such rotors is presented by the authors, as well as by several discussers. Parkinson and Bishop [104] considered the effect of residual vibration after balancing due to the higher, unbalanced modes. A method is proposed by which an averaging technique is used, together with an additional balancing plane, to reduce this residual vibration.

Bishop and Parkinson [19] presented a further discussion of the effect of gravity sag on rotor whirl, with particular emphasis on the second-order whirl found in rotors with dissimilar lateral shaft stiffness (e.g., two-pole generators). The stability of such a rotor was also considered. Bishop and Mahalingam [16] discussed the results of a series of tests with a rotor of this type and extended the discussion of reference [19]. Parkinson [102] then proposed a technique for balancing such rotors, as an extension of *modal balancing*. Parkinson [103] also considered the effect of anisotropic bearing properties, in terms of the non-rotating coordinate system, on the response of a uniform rotor. The effect of the degree of nonsymmetry was discussed and a further extension of *modal balancing* to handle such cases was proposed.

Parkinson [100] summarized the work done by this group in the areas of unbalance response, stability, and balancing of flexible rotors, and presented a complete outline of the *modal balancing* technique. In 1972, Bishop and Parkinson [18] restated the case for N plane *modal balancing* and claimed that the additional expense of (N+2) plane balancing is not justified by the resultant insignificant decrease in residual vibration. Discussions by Kellenberger and Federn disputed this point. Fawzy and Bishop [47], and Parkinson [101], discussed the modal interpretation of measured rotor vibration and bearing effects, particularly as related to the application to *modal balancing* (1976).

A number of other researchers have also published in the general area of modal balancing. Federn [48] and Dimentberg [40] published state-of-the-art reviews of modal balancing in the mid-1960s. Moore and Dodd [93] discussed the practical application of modal balancing to turbine-generator rotors in 1964. Actual experimental results were presented with particular emphasis on a rotor which exhibited significant mixed mode behavior. Kushue and Shlyaktin [76] proposed a modal balancing procedure, coupled with a method for analytically predicting mode shapes. As in most modal balancing techniques (Parkinson's method being a notable exception), the requirement for analytically predicted critical speeds and mode shapes is an integral component of the balancing method. This is generally considered to be a disadvantage, since the typical error associated with such analytical results generally has a significant detrimental effect on the balancing results.

Also in the mid-1960s, Kellenberger [69] proposed a balancing technique similar to the modal balancing techniques discussed above, with one notable exception. He insisted that it is advantageous to balance a flexible rotor initially as a rigid rotor, prior to performing flexible rotor balancing. Then, in order to maintain the rigid rotor balance during flexible rotor balancing it is necessary to use two additional balancing planes; thus the (N+2) plane method mentioned above. In 1972, Kellenberger [71] continued his argument for (N+2) plane balancing, with a discussion by Bishop and Parkinson. Kellenberger's procedure requires analytical prediction of the rotor mode shapes. A further discussion of the N plane method and the (N+2) plane method may be found in Chapter 5. Later (1976), Kellenberger, Weber and Meyer [70] presented a further discussion of N and (N+2) plane modal balancing, as well as influence coefficient balancing, as applied to large steam turbines. They also described facilities for overspeed testing of flexible, large steam turbine rotors.

Miwa [92] also made a case for (N+2) plane modal balancing which he supported with numerical examples. He showed that the method was independent of bearing stiffness and, with Shimada and Nakai [118], extended this result to systems with viscous damping at the bearings.

Additional contributions to modal balancing have been made by Church and Plunkett [22], and Hundal and Harker [65]. Neither of these papers presented anything particularly new, although the use of shake-test measurements for the determination of mode shapes in the former paper is interesting. This is, of course, helpful only for a rotor with very well-defined and easily modeled bearing properties.

In 1976, Lund [84] proposed a variant on modal balancing intended to expand the range of applicability and accelerate the procedure of modal balancing. This acceleration is achieved by reducing the number of trial mass runs to, at most, one per bearing. While analytical predictions are still required, they involve only the free-free mode shapes of the shaft. These can generally be predicted with a somewhat higher reliability than can the bearing support properties. No tests results were presented to verify this technique, though it was stated that a test program was being planned.

Gasch and Drechsler [50] proposed a modal balancing method which requires no trial masses and no knowledge of modal damping or stiffness. Analytically produced mode shapes and modal mass values, together with empirical data taken at a number of speeds approaching a critical speed, are used to identify modal damping and unbalance, using a least squares approach. This procedure would be repeated for each critical speed in the operating speed range. Gnielka [57] extended this method for multi-bearing systems with initial shaft bow. Tests were conducted to verify the effectiveness of this approach using both rolling-element and fluid-film bearings.

Saito and Azawa [117] proposed a variation of modal balancing which provides for the existence of complex modal response due to the presence of viscous damping. This method requires the calculation of *modal exciting factors*, or sensitivities, based on predicted complex eigenvalues and eigenvectors. Modal corrections are then calculated using predicted modal constraints and measured response, without the need for trial mass data. As a result, the required number of test runs is reduced. This method includes an unspecified procedure to account for inaccuracies in the predicted critical speeds. However, there is no apparent way to account for errors in the modal damping ratios, which could substantially affect the predicted modal sensitivities. Also, while this method provides for *light* damping, it appears that the traditional modal balancing assumption of planar mode shapes is still invoked, as all of the examples presented result in planar modal mass sets.

Morton [94] attempted to avoid the difficult issue of predicting rotor support properties by defining the response to be a combination of free modes and rigidly supported modes, which can generally be predicted with reasonable accuracy. Morton's premise is that, with accurate free and rigidly supported modal information, the actual, flexibly supported, modal components of the response can be deduced without knowledge of the bearing properties. The modal unbalance contributions can subsequently be deduced from this response information. One of the procedures for separating the overall rotor response into the free and rigidly supported contributions relies on the fact that the response at the bearings must be due only to the free modes. Using this fact, it is a simple matter to extract as many free mode magnitudes as there are bearings, assuming no other free modes contribute. This approach is clearly limited to shafts operating above a relatively few free modes. Morton also derived a set of equations relating the free and rigidly supported modal response to the unbalance in the shaft. The rigidly supported response terms, in these equations, fall out when the running speed is equal to a rigidly supported natural frequency, further simplifying the calculation of unbalances. He identified some significant limitations to this method, but indicated that there are many situations for which this method would be appropriate in its proposed form, or could be used as an enhancement to one of the more traditional modal balancing procedures.

## Influence Coefficient Balancing Literature

Some of the earliest work in rotor balancing was performed by Thearle [133] in 1934. He developed a two-plane balancing method, particularly suited to field balancing, which can be considered to be the forerunner of influence coefficient balancing. Thearle's method was essentially a two-plane, two-sensor, single-speed, exact-point influence coefficient balancing procedure. The calculations required, even for this simple case, were fairly involved. Consequently, no effort was made toward generalizing the influence coefficient procedure, which requires much more complex calculations, before digital computers became generally available. While Thearle made no specific mention of flexible rotors, it appears that his method would have been suitable for some simple, flexible rotors.

A few years after Thearle's paper, Baker [11] presented a slight generalization of Thearle's method which was essentially a limited, exact point influence coefficient procedure. Baker also suggested some modifications to this method which, while presumably suitable for rigid rotors, are not at all appropriate for flexible rotors. He also discussed the designs of two balancing machines for rigid rotors.

The influence coefficient procedure, as such, was apparently first introduced by El-hadi [44] in a very basic form, some years later (early 1960s). This procedure was proposed in a more formalized and expanded form by Goodman [60] in 1964. This paper is generally considered to represent the introduction of practical influence coefficient balancing. Goodman extended the basic influence coefficient procedure to include least-squares and weighted least-squares solutions. In an interesting discussion to this paper, Heymann [60] described a similar balancing procedure which was in use elsewhere and discussed some of the practical considerations involved in the use of such a procedure.

Lund added some enhancement to the influence coefficient balancing method in a computer program reported by Rieger [113] in 1967. In this reference Rieger also presented the results of an analytical study of the effectiveness of the influence coefficient method which were, in general, very favorable (also reported in [112]). In 1972, Lund and Tonneson [87, 88] reported the details of Lund's influence coefficient procedure along with the results of an experimental investigation which verified its effectiveness. A continuation of this work was reported by Tonneson [137, 138].

At about the same time, LeGrow [79] proposed an alternative balancing method, identical to exact-point influence coefficient balancing except that the influence coefficients are determined analytically. This produces the obvious advantage of eliminating the trial mass runs. However, analytical predictions of this type are prone to significant error which can have a considerable detrimental effect on the results, and efficiency, of the balancing procedure. Also LeGrow's claim that an exact point solution is always preferable to a least-squares solution is clearly open to dispute.

Tang and Trumpler [124], and Yanabe and Tamara [144] proposed different balancing procedures which are both vaguely related to influence coefficient balancing and are both strongly associated with analytical rotor response prediction procedures. Tang and Trumpler proposed a method whereby *disk sensitivities* are predicted analytically and the balancing is accomplished by adjusting the angular orientation of the disks. Yanabe and Tamara used an analytical procedure for determining a matrix of *stiffness coefficients* which are subsequently used in balancing calculations. Apparently, neither of these methods has seen any further development.

Over a period of about ten years (1970s), a large number of experimental investigations were performed by Tessarzik and his co-workers at Mechanical Technology Incorporated (MTI) to evaluate Lund's influence coefficient procedure and computer program. These test programs, as well as several analytical studies, were intended to determine the effectiveness and practicality of influence coefficient balancing for a variety of different rotor types, sizes and operating conditions. Based on the experience gained in these studies, a number of modifications and extensions were made to Lund's program.

In the first of these tests, Tessarzik [125] used the exact point-speed influence coefficient procedure (i.e. the total number of readings equal to the number of planes) to balance a three-mass rotor through two rigid and one flexural critical speed. The results of these tests, along with a discussion of the practical aspects of applying this balancing method, were presented by Tessarzik, Badgley and Anderson [129]. A second series of tests was performed by Tessarzik [127], using the same test rig, to evaluate the least-squares influence coefficient balancing method. These tests were expanded to include other rotor configurations (i.e. various disk masses) and more difficult initial unbalance conditions. The results of these tests were compared with those from the exact point tests to evaluate the relative merits of the two procedures [128]. In general, the least-squares method was found to be superior. Tessarzik [126] subsequently performed a third series of tests with this test rig. The rotor was modified to provide four flexural critical speeds in the operating speed range. However, two of these modes were heavily damped and did not require balancing. These tests were conducted for a number of test rig operating and unbalance conditions, including both rigid and flexible bearing pedestals. In the case of the flexible pedestals, the supports were intentionally designed to be anisotropic, with softer horizontal than vertical supports, resulting in rather extreme elliptical orbits [130]. In all cases, the balancing procedure was quite successful.

This same test rig was later modified by Smalley, Tessarzik and Badgley [121] to provide substantial nonsymmetry of the lateral rotor stiffness. While the principal purpose of these tests was to investigate the stability of such a rotor, the balancing of the nonsymmetric rotor proved to be quite interesting. The least squares influence coefficient balancing procedure verified earlier by Tessarzik [126] was used very successfully for this rotor [122]. This balancing procedure was incorporated into a minicomputer-based automatic data acquisition and computation system, as described in an appendix to reference [122]. The software used in this balancing system, which is an adaptation of Lund's original program, was documented by Darlow and Fanuele [30]. In another appendix to reference [122], Lund provided an interesting discussion on the optimization of balancing plane and sensor locations for influence coefficient balancing.

Much of the balancing work performed at MTI has concentrated on particular rotor types which have tendencies to be very flexible. Two of these rotor types are gas turbine engine power turbine shafts and high-speed power transmission shafts. The investigations into the application of influence coefficient balancing to gas turbine engines began with an analytical feasibility study by Rieger and Badgley [115]. This work was subsequently extended and generalized by Badgley and Rieger [6]. Badgley [2, 4] discussed some of the practical implications of balancing gas turbine rotors. Badgley et al. [8] and Badgley [3, 5] also presented similar discussions for a more general class of flexible rotors. Cundiff, Badgley and Reddecliff described a test rig designed to simulate a gas turbine engine power turbine shaft [23], and the balancing of this simulator through two flexural critical speeds using influence coefficient balancing [24]. A related application of influence coefficient balancing involved the successful simultaneous balancing of a pair of marine steam turbines by Tessarzik, Darlow and Badgley [131].

A similar investigation of high-speed power transmission shafts began with an analytical study by Badgley and Tessarzik [9]. In this paper, the feasibility of designing, manufacturing and balancing a supercritical power transmission shaft was demonstrated, at least theoretically. This followed earlier work on high-speed power transmission shafts by Baier and Mack [10]. The work of reference [9] was later extended and generalized by Badgley, Smalley and Fleming [7]. A scale model test rig (approximately 10 meters in length) of such a power transmission shaft was subsequently designed and built, as reported by Darlow and Smalley [32]. Several series of balancing tests were conducted with this test rig, some of which are described in detail in Chapter 8. These tests included influence coefficient balancing of a torsionally unloaded shaft, as described by Darlow and Smalley [33]; and similar tests with a shaft operating under a substantial torque load, as reported by Darlow, Smalley and Fleming [34]. These tests involved influence coefficient balancing of a test shaft through three active, but four total, flexural critical speeds.

Another area of investigation at MTI involved the use of high-power lasers for removing metal for rotor balancing. Applying lasers in rotor balancing in this way serves the dual purpose of fully automating the balancing procedure and permitting the application of compensating unbalance without stopping the rotor. The second of these points is particularly advantageous for rotors that require excessive time for startup or shutdown, such as large steam turbines. The use of a laser for single plane balancing of simple rigid rotors has been demonstrated by Popick and Roberts [110], Dobson [41], and Damon [25]. A study was subsequently conducted by Tessarzik and Fleming [132] to evaluate the feasibility of incorporating laser metal removal into the multiplane influence coefficient balancing process. The results of this study were quite encouraging. Consequently, the test rig of reference [126] was modified to include a microprocessor-controlled laser metal removal system, as reported by DeMuth, Fleming and Rio [37]. This system was successfully used for fully automatic two-plane balancing.

Analytical and experimental investigations involving influence coefficient balancing have been conducted by a great many other researchers during the past decades. While the majority of these studies have not involved the addition of anything new to the concept or practice of influence coefficient balancing, a few of them have involved some interesting and innovative proposals.

Little and Pilkey [82] proposed a modified influence coefficient method based on a linear programming approach, in 1975. This method requires the number of balancing planes to exceed the number of vibration readings. Thus, an infinite number of solutions are possible. The optimum solution is chosen such that an objective function is minimized. This objective function is generally taken to be the residual vibration at a speed, or operating conditions, at which observations cannot be made. Such an objective function requires the use of analytically determined influence coefficients. The ordinary influence coefficient equations are then treated as constraints. An additional constraint may be added to require the sizes of the correction masses to fall within an acceptable range, thus preventing the calculation of excessively large, or small, correction masses. Even in cases where it is not possible to provide analytical influence coefficients for the objective function, this method still has the advantage of permitting restrictions on the sizes of the correction masses. This balancing method was described in more detail by Little [81], and verified with some interesting analytical examples. However, no experimental results were reported. A similar balancing method was described by Gleizer [56] for use in gas turbine engine flexible rotor balancing. Pilkey and Bailey [108] proposed a "constrained method" of optimization for minimizing residual vibration while constraining correction mass size. This is also a linear programming approach. Pilkey, Bailey and Smith [109] then extended this approach to include optimization of balancing plane axial location, using linear interpolation of the influence coefficients between the existing balancing planes.

At about the same time, Nakai and Miwa [96] and Woomer and Pilkey [143] proposed replacing the linear programming approach with a quadratic programming approach. Again, the object is to minimize the residual vibration while constraining the size of the corrections. Nakai and Miwa also suggested optimizing correction plane selection, presumably from among a set of existing planes for which influence coefficients are available. They verified their method experimentally while Woomer and Pilkey verified their approach numerically. Lee and Kakad [78] extended the linear, or quadratic, programming problem to a general, nonlinear optimization problem. They applied a heuristic combinatorial solution algorithm, which they claimed to be very efficient, although it is still much more computationally intensive than is linear programming. Wang [142] proposed the use of a general optimization approach with correction mass size and residual response constraints.

Nicholas, Gunter and Allaire studied the unbalance response [97] and balancing [98] of a single mass flexible rotor with residual shaft bow. They determined that the optimum results were obtained by reducing the total rotor amplitude, including shaft bow, to zero at the critical speed, and to levels less than or equal to the bow at other speeds. However, this requires prior knowledge of the critical speed, which is not often known with sufficient accuracy. It was shown that an acceptable balance could be achieved by balancing the rotor to the residual bow amplitude at a speed, or speeds, in the general vicinity of the critical speed. Balancing the total rotor amplitude to zero at a speed away from the critical speed was shown by the authors to give totally unacceptable results.

In 1976, Larsson [77] presented a statistical procedure for calculating influence coefficients from multiple sets of trial mass data. He used a linear regression analysis to determine the influence coefficients that result in a *best fit* to the experimental data. The algorithm was constructed in such a way that data from runs with groups of trial and/or correction masses may also be used. In addition to the influence coefficients, Larsson's algorithm also produces a coefficient of multiple determination and a coefficient of reliability. The former of these coefficients provides a measure of the correlation between the influence coefficients and the trial mass data, while the latter provides a measure of the reliability of the calculated correction masses. Larsson discussed the results of tests using two productions runs of two types of rotors; 31 of one rotor type and 2 of the other. In these tests, the influence coefficients were calculated, and continually improved, through the use of his algorithm. The use of these influence coefficients aided substantially in the balancing of these rotors and, toward the end of a production run, permitted the use of fewer, or in some cases no, trial mass runs.

Drechsler [43] has proposed a similar procedure for applying the tools of statistical analysis to the calculation of influence coefficients. However, his discussion was concerned primarily with feasibility and a general approach, and lacked the detail and development of Larsson's presentation. Drechsler did not present any experimental verification.

In 1977, Bigret, Curami, Frigeri and Macchi [12] described a modified influence coefficient balancing method adapted for use with an in-field computer. This balancing method is the standard weighted least-squares influence coefficient method modified to use a generalized weighting of particular speeds or sensors. The in-field computer system used relied on manual input of vibration data by means of a Teletype. This balancing system was experimentally verified, with good success, as described in the paper.

Fujisawa, Shiohata, Sato, Imai and Shoyama [49] reported good results from an experimental investigation in which the least-squares influence coefficient method was used for balancing a multispan, flexible rotor. Although there was nothing innovative about their balancing method, this does appear to be the first publication of actual balancing results for multispan rotors using an influence coefficient balancing procedure. Since such rotors are of considerable practical importance (e.g., modern steam turbine-generator sets), this paper is worthy of note. Shiohata and Fujisawa [119] subsequently applied a correction mass magnitude constraint to the least squares method and demonstrated it using the same multispan rotor. With Sato [120], they also developed a procedure for axially locating an unbalance in a flexible rotor from journal vibrations. However, the practicality of this procedure was not well established, as no indication was given that this could actually be used for balancing. In addition, it appears that this procedure is predicated on knowing the angular location of the unbalance, which is not realistic.

Darlow [27] developed a procedure for identifying and eliminating non-independent balance planes in influence coefficient balancing. Used as a part of the influence coefficient balancing method, this procedure results in a reduced-rank, well-conditioned influence coefficient matrix which is then used for calculating correction masses in the remaining, independent balancing planes. This procedure is described in detail in Chapter 6.

A low-speed, flexible rotor balancing method was proposed by Giordano and Zorzi [52]. The object here is, in a sense, the reverse of modal balancing. In modal balancing the intention is to take data near to a critical speed so that no other modes produce significant response. However, the intent in the low-speed method is to take data between two critical speeds in such a way that both modes contribute significantly. The *low speed* feature is achieved by reducing the bearing stiffnesses such that operation between the first two flexural critical speeds can be achieved at speeds below what would normally be the first critical speed. Once the operating conditions are selected, influence coefficient balancing is used. While the definition of this method in [52] is stated using more general terms such as *away from critical speeds*, the examples presented indicate that in order to achieve the combination of *sensitivity* and *plane separation* required, balancing must take place between the first two flexural critical speeds. It is also not clear whether this method can be used for rotors which must traverse more than two flexural critical speeds. Despite its limitations, there are certainly situations for which this method can be very advantageous.

Recently, Euler and Speckhart [45], and Everett [46], have published descriptions of procedures for influence coefficient balancing without the need for phase measurements. These are both basically analytical, multiplane extensions of the classic four-run, polar plot method. With modern instrumentation, the measurement of phase is neither as difficult, nor as expensive, as it was only a decade ago. Thus, it would not generally be justified to conduct the additional trial runs which are required for either of these methods. A paper by Darlow [28] and a trade journal article by Hogan [64] describe portable balancing systems which are indicative of the sophisticated instrumentation now available at a modest cost.

## Literature on Combined Modal/Influence Coefficient Balancing Methods and Comparative Studies

Rieger [114] provided the earliest overview which considered both modal and influence coefficient balancing. While this reference appeared earlier (1973) than most of the influence coefficient balancing literature, Rieger was clearly making a case for the superiority of influence coefficient balancing over modal balancing.

Giers [51] evaluated the N and (N+2) plane modal balancing methods, as well as influence coefficient balancing from the viewpoint of practical application, which drew strong discussion from proponents of the various methods. Of particular interest is a discussion of the practical aspects of data acquisition in rotor balancing. Giers also stated that an improved method could be devised from a combination of modal and influence coefficient balancing, but did not indicate how this would be accomplished.

Drechsler [42] proposed a balancing method which is, to a certain extent, a combination of modal and influence coefficient balancing. He simply applied the standard influence coefficient equations and added a weighting matrix designed to account for the analytically predicted mode shapes of the rotor. Individual trial masses are used throughout this balancing procedure. Thus, while this method is not prone to the potential numerical problems of influence coefficient balancing, no attempt is made to deal with the inherent difficulty of obtaining reasonable trial mass data at speeds above one or more of the lowest flexural critical speeds for a lightly damped system. Also, the use of analytically determined mode shapes is a distinct disadvantage borrowed from modal balancing. Drechsler presented a numerical example to verify his balancing method. However, no experimental verification was provided.

Black and Nuttal [20] considered modal balancing of rotors with more than *light* damping. They showed that this could be done using a bi-normal orthogonalization, but would require two planes per mode. If, on the other hand, the damping were *appreciable but small*, the use of the extra balancing planes would result in ill-conditioned equations. They concluded that it would probably be preferable to use a direct observation approach, such as influence coefficient balancing, "combined with a clear analytical understanding of the shaft modal structure", but did not provide any details concerning how this would be accomplished.

Maxwell and Sanderson [90] illustrated the need to subtract static runout from uncorrected rotor data. They used influence coefficient balancing, described as an extension of Thearle's method. They enhanced this approach by extracting modal components of the response data to generate modally constrained connections, but did not explain how this extraction was done.

Bulanowski [21] provided a brief background in modal and influence coefficient balancing. He stated that influence coefficient balancing is complimentary to modal balancing and they should be used together. He appears to use modal information only for plane selection prior to standard influence coefficient balancing.

Iwatsubo [66] described modal and influence coefficient balancing saying that modal balancing is really a special case of influence coefficient balancing. He compared modal and influence coefficient balancing using numerical examples. He showed that influence coefficient balancing is more sensitive to measurement error than is modal balancing. He indicated that a combined method would be preferable, but gave no details as to how this would be accomplished.

In a series of publications [29, 35, 36, 105, 106, 107], Parkinson, Darlow, Smalley and Badgley described the Unified Balancing Approach. This method uses empirically determined modal characteristics to enhance, and substantially modify, influence coefficient balancing. Theoretical basis, practical procedure and experimental verification results are presented in these publications. A detailed analytical development and practical procedure for the Unified Balancing Approach are presented in Chapter 7.

Darlow and Smalley [31] described how the principle of reciprocity can be used to reduce the number of trial mass runs required when using either influence coefficient balancing or the Unified Balancing Approach. Presumably, this principle could also be applied to some of the more empirical modal balancing approaches. This particular application of the principle of reciprocity is discussed in more detail in Chapter 9.

# Chapter 4
# Rotor Balancing Methods and Instrumentation

Rotor balancing methods may be separated into two categories: rigid rotor balancing and flexible rotor balancing. These generic titles refer to whether or not rotor flexibility is taken into account. That is, flexible rotor balancing procedures take into account rotor flexibility, while rigid rotor balancing procedures do not. Many flexible rotor balancing methods are suitable for balancing rigid rotors. However, rigid rotor balancing methods are, in general, not effective for balancing flexible rotors. Unfortunately, in practice, the use of rigid rotor balancing methods with flexible rotors is not unusual. The result is, at best, an extremely inefficient and costly balancing process and, at worst, a presumably well-balanced rotor which, at its operating speed, is actually very poorly balanced and potentially destructive. Specific rigid and flexible rotor balancing methods are discussed in considerable detail in this and the following several chapters.

Field balancing is a term which appears repeatedly in the literature and deserves some explanation here. This term refers to a situation in which a balancing method is applied on-site, generally using portable instrumentation. It is not a particular type of balancing procedure or classification. The rotor being field balanced may be either rigid or flexible. Generally, the procedures most commonly used for field balancing utilize only a single balancing plane [67] or multiple balancing planes in a trial-and-error fashion, such as the polar plot methods described below. In either case, balancing of flexible rotors is, at best, inefficient and more likely, ineffective. Some efforts toward development and promotion of systematic flexible rotor balancing procedures suitable for field balancing have been reported [4, 128], primarily in the fields of influence coefficient balancing and the Unified Balancing Approach. The emergence of the portable microcomputer as a field instrument has made these developments possible.

## Rigid Rotor Balancing

The classification of rigid rotor balancing encompasses a number of balancing procedures whose common feature is the assumption that the rotor being balanced does not elastically deform at any speed up to its maximum design speed. If the rotor being balanced does not satisfy this assumption, rigid rotor balancing is not adequate and flexible rotor balancing, as described in the following chapters, must be used.

Rigid rotor balancing is divided into two distinct classifications commonly referred to as *static* balancing and *dynamic* balancing. Static balancing refers to single-plane balancing, which is appropriate for rigid rotors in which the mass is concentrated in a single narrow disk. Thus, the removal of any mass eccentricity in the disk effectively balances the rotor as a whole. The single-degree-of-freedom rotor model illustrated in Figure 4.1 is a classic representation of such a rotor. Examples of actual rotors appropriate for static balancing include gyroscopes and bull gears on short, rigid shafts. In general, many types of single disk straddle mounted rotors with short bearing spans may be effectively balanced by static balancing.

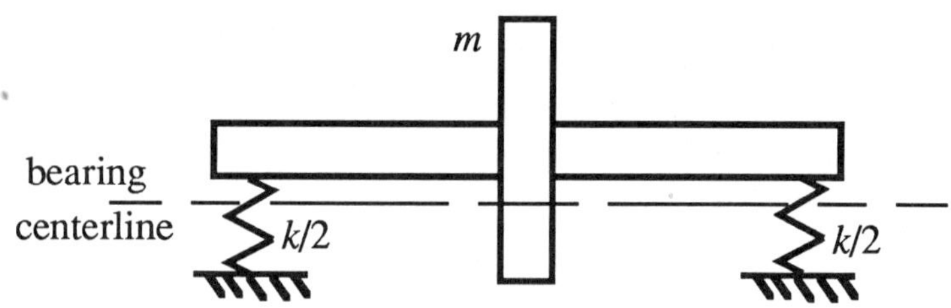


**Figure 4.1** Single Degree-of-Freedom Rigid Rotor Model

Dynamic balancing refers to two-plane balancing. This is somewhat of a misnomer in that it implies that the balancing takes into account all of the dynamics of the rotor, including rotor deformation. This, of course, is not the case, as dynamic balancing is limited in application to rigid rotors. However, as discussed in Chapter 2, any rigid rotor can be completely balanced in two planes, regardless of whether it is rigidly or flexibly mounted. This conclusion is based on the general rigid rotor model illustrated in Figure 4.2 and the accompanying analysis.

In particular, any general unbalance distribution may be represented as a linear combination of an $a$ and a $\theta_a$ and some arbitrary unbalance distribution which is completely self-cancelling. A self-cancelling unbalance distribution is one which results in no forces or moments being transmitted to the bearings. That is, the resulting unbalance forces satisfy

$$\int_0^l u_c(z)dz = 0 \tag{4.1}$$

$$\int_0^l z u_c(z)dz = 0 \tag{4.2}$$

where $u_c(z)$ represents the self-cancelling unbalance distribution; $z$ is measured in the axial direction; and $l$ is the length of the rotor. An example of a general unbalance distribution and its components is given in Figure 4.3. Although the unbalance distribution components in this figure are shown to exist in a single diametral plane, or longitudinal cross-section, this is done for clarity only. In general, these components are found to exist in diametral planes which are rotated with respect to one another.

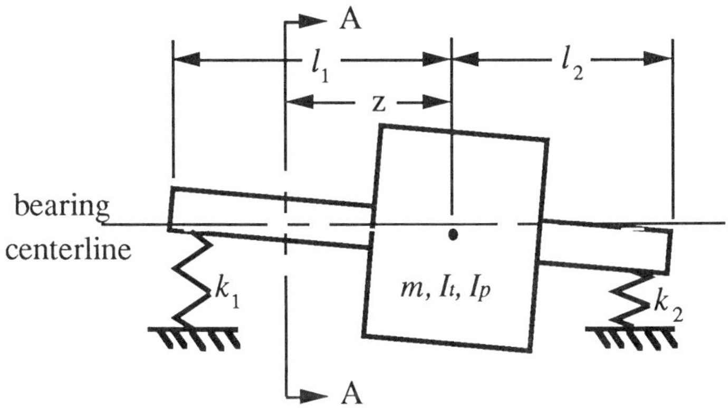


**Figure 4.2** General Rigid Rotor Model

The $a$ component of unbalance may be canceled with a single correction mass and a pair of correction masses are required to cancel the $\theta_a$ component of unbalance. However, no correction is required for the self-canceling component. Since the axial location of the correction mass for the $a$ component is arbitrary, this mass may be located in the same diametral plane as one of the correction masses for the $\theta_a$ component. Since two correction masses located in the same diametral plane can always be replaced by the single mass which represents their vector sum, in rotating coordinates, it becomes apparent that a rigid rotor can always be balanced in, at most, two planes. Thus, dynamic balancing is always sufficient for a rigid rotor. Note that if the correction for $a$

is not located in the same transverse plane as the center of gravity, $\theta_a$ is modified.

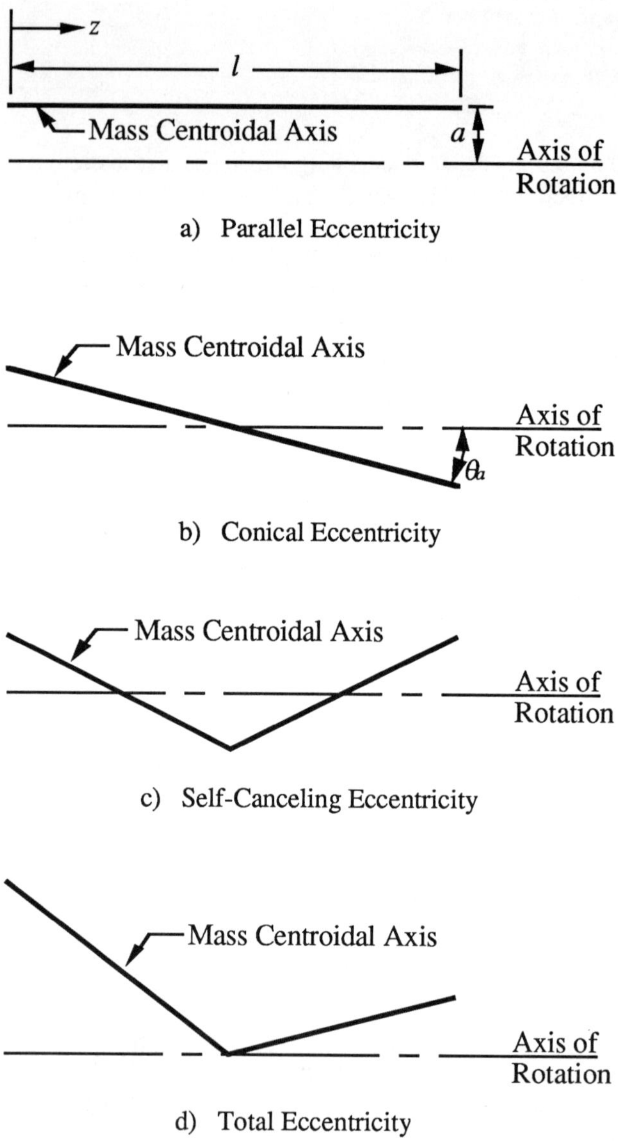


**Figure 4.3** Distribution of Mass Centroidal Axis Eccentricity and the Effective Components in Terms of Rigid Rotor Response

The differences among the various existing machines and methods for rigid rotor balancing are more closely related to implementation than to theory. Specific machines and methods for both static and dynamic balancing are discussed in some detail.

One of the oldest and most commonly used methods for static balancing involves the use of horizontal knife-edges. The rotor is set on the knife-edges and is permitted to rotate until the heavy side faces down. Some mass is then attached to the light side, or top, of the rotor and the process is repeated until the rotor has no preferential orientation. A static balancing machine which uses a similar procedure is the bubble-type automobile tire balancer. One of the primary drawbacks of this machine is that the center of the rotation of the axle relative to the wheel must be exactly the same as the pivot point on the tire balancer. Any disparity will have a detrimental effect on the balance condition of the tire, wheel and axle assembly.

A somewhat more systematic method of static balancing, or single-plane balancing in general, which is particularly common in field balancing, is referred to as the *polar plot* method. Obviously, the knife-edge procedure is not suited to field balancing. In the *polar plot* method, rotor vibration is first measured, generally at a low speed, for the initial unbalanced condition, referred to as *uncorrected rotor data.* Then one or more trial mass runs are conducted where a trial mass of known size is installed in the rotor at a predetermined angular location and rotor vibration is measured at the same speed as for the uncorrected rotor data. The resulting rotor vibration data is referred to as *trial mass data.* Using this rotor vibration data and fairly simple polar plotting techniques, the appropriate unbalance compensation, or correction mass, is calculated and installed in the rotor. If the rotor vibration is not sufficiently reduced (a not uncommon occurrence), the entire procedure is repeated to refine the correction mass.

This polar plot calculation procedure is illustrated by the example of Figure 4.4. In this plot the vector labeled $\hat{U}$ represents the *uncorrected rotor data* and the vector labeled $\hat{T}$ represents the *trial mass data.* The vector resulting from the subtraction of vector $\hat{U}$ from vector $\hat{T}$ is labeled $\hat{D}$ and represents the specific effect of the trial mass on the rotor response. It is then simply a matter of modifying the trial mass magnitude and angle so that the resulting vector $\hat{D}$ exactly cancels the current rotor vibration vector. If, as in most cases, the original trial mass has been removed, the current rotor vibration vector refers to the vector $\hat{U}$. However, in some cases it is necessary to grind or otherwise permanently install the trial mass. In such a case, the current rotor vibration vector refers to the vector $\hat{T}$. In either case, the assumptions of linearity and superposition tell us that modifying the trial mass magnitude and/or angle will have a corresponding effect on the vector $\hat{D}$. For example, the vector $\hat{U}$ in Figure 4.4 has a magnitude of 1.5 and an angle of 45 degrees., while the vector $\hat{T}$ has a magnitude of 2.0 and an angle of 120 degrees. Then the vector $\hat{D}$, which is the difference between $\hat{T}$ and $\hat{U}$, has a magnitude of 2.2 and an angle of 162 degrees. By moving the vector $\hat{D}$ through an angle of 63 degrees and reducing its magnitude by 30 percent, it will become the negative of the vector $\hat{U}$. Consequently, assuming the trial mass is not permanent, the unbalance response can be eliminated by moving the trial mass through the same angle of 63 degrees and reducing its magnitude by the same 30 percent. Thus, if a one gram trial mass was used at zero degrees, the correction mass would be 0.7 grams at 63 degrees. On the other hand, if the trial mass is permanent, the vector $\hat{D}$ would be modified to cancel the vector $\hat{T}$. This would involve an angle shift of 138 degrees and a 10 percent reduction in magnitude. The corresponding correction mass would then be 0.9 grams at 138 degrees. The direction, relative to the direction of rotation, which corresponds to a positive angle for a trial or correction mass, depends upon the angle convention used for the measurement of vibration data. If the vibration data is measured using a leading phase angle, a positive angle for a mass location is measured in the same direction as rotation. Conversely, the use of a lagging phase angle corresponds to positive mass angle measurement in a direction opposite that of rotation. This *polar plot* balancing procedure is, in essence, a very simplified, single-plane influence coefficient balancing operation.

If vibration data can be measured in terms of amplitude and phase angle, relative to an angular reference fixed in the rotor, only a single trial mass run is required for a correction mass calculation, as discussed above. However, if only vibration amplitude is available, then a minimum of two trial mass runs are required. In general, three trial mass runs are used in this situation since the use of only two such runs permits specification of the diametral plane for the correction mass, but not the direction. That is, there are two possible choices which are 180 degrees out of phase. The use of three trial mass runs allows the complete specification of the correction mass. The relative ease with which accurate phase measurements can be made using current instrumentation has rendered this method archaic and unnecessarily time consuming. However, for the interested reader, more specific information regarding the implementation of this *four run* method may be found in reference [140].

In general, dynamic balancing machines and methods are perfectly adequate for static balancing, although static balancing machines and methods are certainly not adequate for rotors which require dynamic balancing. Many ingenious methods were developed for dynamic balancing prior to the advent of modern vibration instrumentation. Most of these methods are no longer used and are not discussed in this book. However, detailed descriptions of some of these methods may be found in references [11, 38, 75, 123, 135]. Of primary interest here are the dynamic balancing machines which are used most often. These machines generally fall into two classifications, commonly referred to as soft-bearing machines and hard-bearing machines, which very aptly describe their differences.

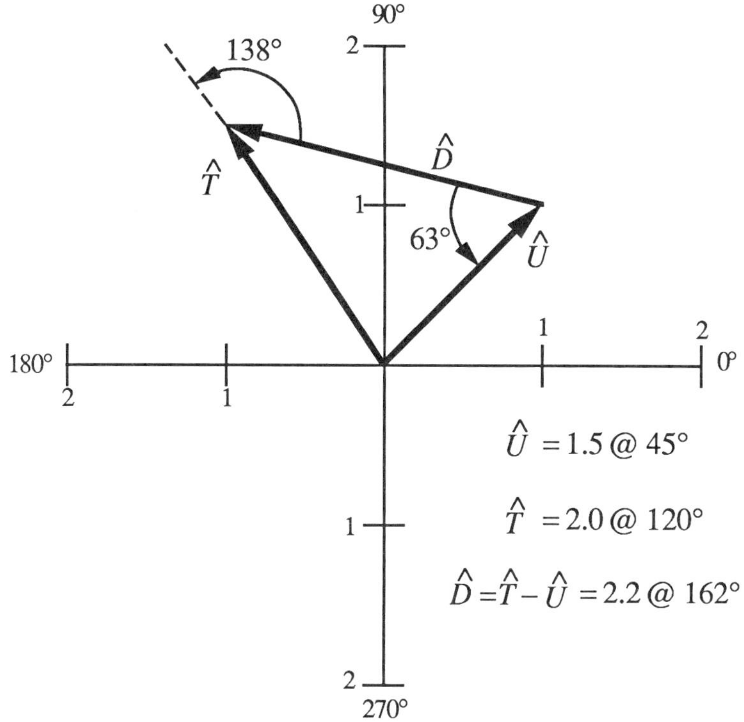


**Figure 4.4** Illustration of Polar Plot Calculation Procedure for Single Plane Rotor Balancing

Soft-bearing balancing machines have rotor supports which are designed to be very soft in the horizontal direction, such that the rigid rotor critical speeds are well below the balancing speed. Thus, the rotor support vibrations are sufficiently large to permit high-resolution displacement measurement. The principal advantage of soft-bearing machines over hard-bearing machines is the excellent sensitivity, particularly important for light rotors. With soft-bearing machines, either trial mass runs are required to *calibrate* the rotor, or else a less systematic trial-and-error procedure must be used. With the development of hard-bearing machines, the popularity of soft-bearing machines has decreased, particularly for production run balancing. This is due largely to the inconvenience of rotor *calibration*, the difficulty of handling massive rotors, and the need for relatively skilled operators for soft-bearing machines. Currently, these machines are used more commonly for high-precision, low-volume balancing requirements and for the balancing of very light rotors.

Hard-bearing balancing machines, introduced more recently than soft-bearing machines, were made possible by the development of sensitive and accurate force measurement transducers, or force gages. In hard-bearing machines, the rotor supports are structurally rigid and are each mounted on one or more force gages. Since the force gages are also very stiff, the balancing speed is far below all rotor critical speeds. Consequently, the rotor vibration, whose displacement amplitudes are very small, and bearing forces, which are measured by the force gages, attain maximum values essentially directly in line (i.e. are exactly in phase) with the rotor unbalance. Also, the bearing forces may be related directly to the mass eccentricity of the rotor, if the rotor mass distribution is known, or to the equivalent discrete unbalances, if the axial locations and radii of the unbalances are known. In general, the axial locations of the unbalances are chosen and the corresponding radii may be easily measured. Thus, the equivalent discrete unbalances can be calculated directly from the bearing forces and the appropriate discrete correction masses can be applied to the rotor.

As discussed above, no more than two balancing planes are required for complete balancing of a rigid rotor. Figure 4.5 illustrates a typical rigid rotor balancing setup on a hard-bearing machine.

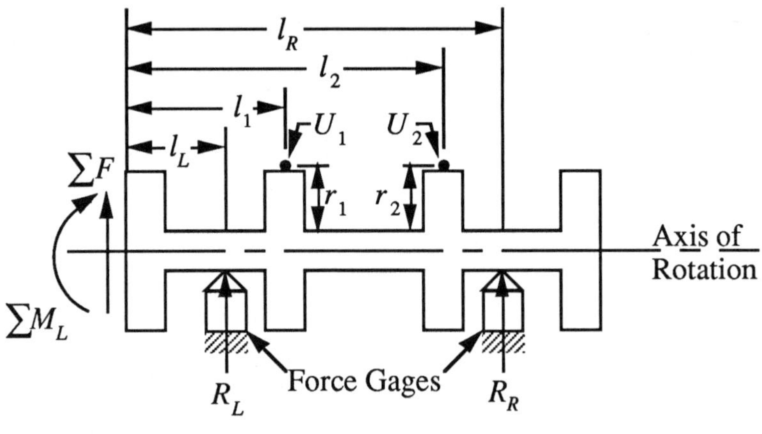


**Figure 4.5** Typical Hard-Bearing Balancing System Setup

For a constant speed of operation, the synchronous components of the bearing forces may be treated as quasi-static forces which are represented by complex numbers, as discussed in Chapter 2. If the moment carrying capability of the rotor supports is assumed to be negligible, which is generally one of the design objectives for the balancing machine, the rotor may be treated as being statically determined. Thus, using simple statics, but with complex valued forces, we have

$$\Sigma F = R_L + U_1 r_1 \omega^2 + U_2 r_2 \omega^2 + R_R = 0 \tag{4.3}$$

$$(\Sigma M)_L = R_L l_L = U_1 r_1 \omega^2 l_1 + U_2 r_2 \omega^2 l_2 + R_R l_R = 0 \tag{4.4}$$

where

$\Sigma F$ represents the sum of the forces on the rotor

$(\Sigma M)_L$ represents the sum of the moments about the left end of the rotor

$R_L, R_R$ are the bearing reactions at the left and right rotor supports, respectively

$U_1, U_2$ are the unknown equivalent discrete unbalances at axial locations 1 and 2 in Figure 4.4.

$r_1, r_2$ are the corresponding radii for application of the correction masses

$l_1, l_2$ are the axial distances from the left end of the rotor to locations 1 and 2 in Figure 4.4

$l_L, l_R$ are the axial distances from the left end of the rotor to the left and right rotor supports, respectively, and

$\omega$ is the speed of rotation in radians per second

Note that the bearing reactions are 180 degrees out of phase with the bearing forces. Equations (4.3) and (4.4) may be solved for $U_1$ and $U_2$ to give

$$U_1 = [R_R(l_R - l_2) - R_L(l_2 - l_L)]/[r_1 \omega^2 (l_2 - l_1)] \tag{4.5}$$

$$U_2 = [R_L(l_1 - l_L) - R_R(l_R - l_1)]/[r_2 \omega^2 (l_2 - l_1)] \tag{4.6}$$

where the result are complex values. The calculation of Equations (4.5) and (4.6) may be accomplished with analog electronics, as was done originally, or with digital components, as in most of the more recent balancing machines. The ability to calculate $U_1$ and $U_2$ simultaneously is often referred to as *plane separation.*

If the correction masses are to be applied by removal of mass from the rotor (e.g., by grinding the rotor), $U_1$ *and* $U_2$ define the amount of mass to be removed from the rotor and the appropriate angular orientations (i.e., the unbalances $U_1$ and $U_2$ are being removed). If, however, the balance corrections are to be made by the addition of mass to the rotor (e.g., by placing set screws in pre-threaded holes), the angular orientations of the corrections must be rotated 180 degrees (i.e. to the opposite side of the rotor) from that defined by $U_1$ and $U_2$ (i.e., the unbalances $U_1$ and $U_2$ are being compensated).

The popularity of hard-bearing balancing machines is currently increasing for generally the same reasons that the popularity of soft-bearing machines is declining. The convenience of one-shot balancing and suitability for production operation are two attractive features of hard-bearing machines. In addition, these machines can be used by relatively unskilled operators and are capable of handling a wide range of rotor sizes.

Typically, the balancing speed for hard-bearing machines is between 600 and 1800 rpm. Generally, higher speeds are used for lighter rotors or where tight balancing tolerances are encountered. The bearing forces, and thus the sensitivity to unbalance, is proportional to the square of the speed of rotation. Therefore, when more sensitivity is required, higher balancing speeds can be used. The balancing speed is, however, limited by the flexibility of the rotor and supports in that it must remain well below the lowest rotor critical speed. Sometimes, particularly in the case of very light rotors, it is not possible to attain sufficient sensitivity at allowable balancing speeds. In this case, it may be necessary to use a soft-bearing balancing machine.

There are certain potential problems associated with the use of rigid rotor balancing machines. These problems are generally a result of improper or inappropriate use of rigid rotor machines and can be avoided if the supervisory engineer is aware of them. First of all, it is essential that rigid rotor machines not be used for balancing rotors which are, in fact, flexible. Second, rotors should have the same centers of rotation on the balancing machine as they do in operation. For example, if a rotor is to be supported by rolling-element bearings, it should, if possible be balanced mounted in these same bearings. If the same center of rotation is not used, substantial unbalance can be introduced which may have a detrimental effect on bearing and rotor life. This is particularly true in cases where rotor disks are balanced individually on arbors, rather than on their shafts. What often happens in this case is that the disk is balanced on an arbor that has very little eccentricity and is then mounted on a shaft with considerably more eccentricity. The result is a large unbalance equal to the mass of the disk times the eccentricity of the shaft on which it is mounted. In cases where it is not possible to duplicate a rotor's operating centers of rotation on a rigid rotor balancing machine, it may be necessary to balance the rotor *in-situ*, completely assembled in place, or *quasi-in-situ*, in a specially constructed balancing rig which simulates the actual machine.

A third source of potential problems with rigid rotor balancing machines occurs when a rotor stack-up must be disassembled after balancing in order to be installed. Considerable rotor unbalance is often introduced in this way, even if care is taken to reassemble the rotor components without changing their angular orientations. This problem can occur with quasi-in-situ balancing, as well as when using rigid rotor balancing machines. One of the principal limitations of these balancing machines is that it must be possible to bring the rotor to the balancing machine. In many cases, particularly with very large rotors, this simply is not possible.

## Flexible Rotor Balancing

A flexible rotor is defined as one which exhibits significant lateral deformation at, or below, its maximum design speed. A survey of the literature on flexible rotor balancing is presented in Chapter 3. The theoretical developments of the three principal flexible rotor balancing procedures are presented in the following three chapters.

The use of rigid rotor balancing procedures for flexible rotors can be a very dangerous practice. Such practices are generally ineffective, and, at the least, inefficient. It is not unusual for the use of rigid rotor balancing to actually aggravate the unbalance condition of a flexible rotor at its operating speed. For example, consider the flexible rotor in Figure 4.6. This is a uniform rotor of length $l$ and radius $r$, supported at its ends by relatively rigid bearings. This rotor has a single discrete unbalance mass $U$ located on the surface of the shaft at an axial distance $0.1l$ from the left end of the rotor. The two correction mass planes, for dynamic balancing, are located at axial distances $0.4l$ and $0.6l$ from the left end of the rotor. The mode shapes for the first two critical speeds are illustrated in Figure 4.7

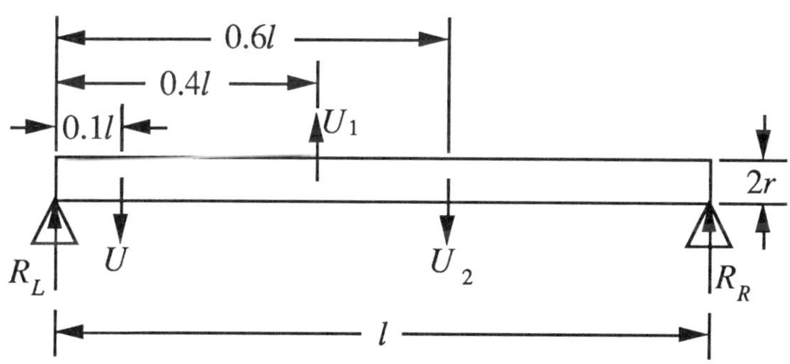


**Figure 4.6** Uniform Flexible Rotor on Rigid Bearings

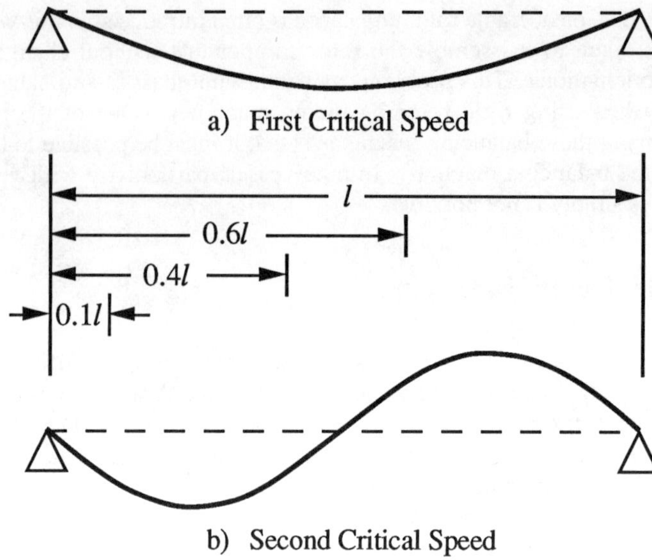


**Figure 4.7** Mode Shapes for Uniform Flexible Rotor on Rigid Bearings

This rotor is balanced as a rigid rotor at rotational speed $\omega$, using dynamic balancing, by measuring the bearing reaction forces, which are $0.9\omega^2 rU$ and $0.2\omega^2 rU$ for $R_L$ and $R_R$, respectively. Then the required correction masses $U_1$ and $U_2$, which are also located on the surface of the shaft, are calculated from Equations (4.5) and (4.6) as $-2.5U$ and $1.5U$, respectively. The amplitude of vibration at, or near, a critical speed, by any means of measurement, is proportional to

$$\sum_{i=1}^{n} U_i r_i x_i \tag{4.7}$$

where the $U_i$ are the $n$ discrete unbalance masses, including the correction masses, and $r_i$ and $x_i$ are the corresponding radii and normalized amplitudes of vibration, respectively. Thus, without rigid rotor balancing, the vibration amplitudes at the first two critical speeds are proportional to $0.32rU$ and $0.59rU$, respectively, where $rU$ is the original unbalance mass moment. After rigid rotor balancing, the corresponding vibration amplitudes will become proportional to $0.64rU$ and $1.76rU$. Thus, for this example, the rigid rotor balancing procedure actually has a detrimental effect on the rotor vibration levels of the flexible rotor. Some flexible rotor balancing procedures require prior rigid rotor balancing. However in general, rigid rotor balancing alone is not adequate for a flexible rotor.

A flexible rotor must, in general, be balanced in the vicinity of all critical speeds below the maximum design speed of the rotor. It is also often necessary to balance a flexible rotor at its operating speed. Occasionally balancing at an overspeed condition is also required. A typical situation where overspeed balancing is desirable is when a mode for a critical speed above the operating speed range has a significant response at the operating speed, indicating a need for balancing. The added balance data sensitivity afforded by balancing at overspeed provides for a more precisely balanced rotor. The ultimate aim of any flexible rotor balancing procedure is to produce acceptable vibration levels, be they bearing reaction forces or displacement amplitudes, at all balancing speeds, not merely at the running speeds. This is because, even for supercritical rotors which operate at a constant speed, it must be possible to safely accelerate to running speed and decelerate to shutdown, going through critical speeds.

## Vibration Measurement for Rotor Balancing

The vibration data used for rotor balancing may come from a variety of sources. One commonly used source of vibration data is dynamic bearing reactions, particularly for rigid rotor balancing. These bearing reactions may be measured in the form of forces (e.g. using force gages), displacements (e.g. using displacement sensors) or, in the case of film bearings, oil pressure (e.g. using pressure sensors). The bearing reactions may also be measured indirectly in the form of bearing pedestal vibrations. The most commonly used device for measuring pedestal vibrations is the velocity pickup, although accelerometers and displacement sensors are also used. The use of pedestal vibration data is most frequently associated with field balancing.

The measurement of dynamic rotor deflection, using displacement sensors, is another important source of balancing data. As rotors have become increasingly flexible, the maintenance of rotor/stator clearance has become a real problem and the use of rotor deflections for balancing data has gained substantial popularity and has, in some cases, become essential. In particular, for extremely flexible rotors mounted on relatively rigid bearings, bearing reaction data may not provide sufficient sensitivity to achieve an acceptable level of balance.

The principal drawback to the use of displacement sensors along a rotor has historically been the general inaccessibility of the rotors in fully assembled rotating machinery, with the possible exception of the ends of the rotors. However, as machinery manufacturers have begun to recognize the need for rotor vibration monitoring for balancing, as well as for general machinery maintenance and diagnostics, they have started installing displacement sensors and other transducers as integral components of their machines. Although the cost of these transducers is negligible when compared to the overall cost of most commercial machinery, a great deal of manufacturer resistance to properly instrumenting rotating machinery remains to be overcome.

When using displacement sensors for measuring rotor deflections, the proper choice of the axial locations of the sensors is very important. If the sensors are all located at, or near, rotor deflection nodes (i.e. points of zero lateral displacement) the rotor may appear to be very well balanced when, in fact, it is not. Such a situation is potentially disastrous. Thus, in general, proper selection and location of vibration sensors is critical to successful and efficient balancing of flexible rotors.

When discussing the acquisition of data for rotor balancing, extraction of the synchronous amplitude and/or phase information is an important consideration. There are two aspects to this data reduction, although they are often handled as a single operation. The first of these is the elimination, through filtering, of any nonsynchronous information in the sensor signal. The second is the indication of the amplitude and/or phase of the remaining synchronous component of the signal. Generally, a keyphasor signal is produced using a sensor to detect a structural discontinuity fixed in the rotor, resulting in a once-per-revolution signal which has a fixed phase relationship with the rotor structure. This signal is used as a frequency reference for the filter and, if phase information is produced, as a phase reference.

Some years ago, the inability to make accurate phase measurements led to the development of balancing techniques which did not require phase information, such as the *four run* polar plot method discussed above. However, more recently it has become possible, and relatively inexpensive, to obtain very accurate amplitude *and* phase information from balancing sensor data. Consequently, use of the more cumbersome, amplitude-only procedures, as well as more recently developed phase-only procedures, can no longer be justified.

# Chapter 5
# Modal Balancing

Modal balancing procedures are characterized by their use of the modal nature of rotor response. In general, modal balancing procedures are highly developed, but are largely intuitive in nature. They seek to balance the rotor, one mode at a time, with a set of masses specifically selected not to disturb previously balanced, lower modes. This chapter includes the analytical background and typical procedures for modal balancing.

## Analytical Development

Neglecting the effect of system damping and lateral anisotropy, which are inherent assumptions of modal balancing, the principal modes of an axially uniform rotor system are found to satisfy the orthogonality condition

$$(\omega_j^2 - \omega_k^2)^2 \int_0^l x_j(z) x_k(z) dz = 0 \tag{5.1}$$

where $\omega_j$ and $\omega_k$ are undamped system resonances, in this case synonymous with critical speeds, for the $j^{\text{th}}$ and $k^{\text{th}}$ rotor whirl modes, respectively; and $x_j(z)$ and $x_k(z)$ are the corresponding planar, though not necessarily coplanar, modal amplitudes [114]. The $Z$-direction is along the rotor axis and the overall length of the rotor is $l$. For $j$ not equal to $k$, Equation (5.1) reduces to

$$\int_0^l x_j(z) x_k(z) dz = 0 \qquad \text{for } j \neq k \tag{5.2}$$

For a more general rotor system with axial nonuniformity, Equations (5.1) and (5.2) may be generalized in the form

$$\int_0^l A\rho(z)\phi_j(z)\phi_k(z)dz = \begin{cases} 0 & \text{for } j \neq k \\ Q & \text{for } j = k \end{cases} \tag{5.3}$$

where the mode shapes $x_j(z)$ and $x_k(z)$ are now represented by the characteristic functions $\phi_j(z)$ and $\phi_k(z)$, respectively [18]. The term $A\rho(z)$ is the mass per unit length, or cross-sectional area times density, of the rotor and $Q$ is the normalizing factor for the characteristic functions. That is, the characteristic functions are normalized such that $Q$, which has dimensions of mass, is constant regardless of the values of $j$ and $k$, so long as they are the same.

Using the property of orthogonality, any fairly well-behaved function $f(z)$, such as mass eccentricity or rotor deflection, can be expressed as an infinite series of the characteristic functions (since there are an infinite number of characteristic functions) by

$$f(z) = \sum_{j=1}^{\infty} f_j\, \phi_j(z) \tag{5.4}$$

where the coefficients $f_j$ are given by

$$f_j = \frac{1}{Q} \int_0^l A\rho(z) f(z)\phi_j(z)dz \tag{5.5}$$

The actual rotor deflection $x(z)$ may be represented by an equation analogous to Equation (5.4), where the coefficients $x_j$ are functions of rotational speed. In general, as the rotational speed approaches the $j$th resonance, $x_j$ becomes substantially larger than the other coefficients. Thus, $x(z)$ becomes nearly equal to $x_j\phi_j(z)$. This is referred to as modal response.

As an example, consider a uniform rotor of length $l$ supported at its ends by radially rigid, but angularly free, bearings. In the absence of damping and gyroscopic effects, the response of this rotor is identical to the lateral vibration of a non-rotating beam with the same dimensions. Neglecting shear deformation, which is negligible for a long slender beam, the moment, $M$, applied to a differential section of the beam, as in Figure 5.1, is related to the curvature of the beam by

$$M = EI\frac{\partial^2 x}{\partial z^2} \tag{5.6}$$

for small deformations. The modulus of elasticity and cross-sectional moments of inertia are represented by $E$ and $I$, respectively, while $z$ is the coordinate measured along the rotor axis and $x$ is the lateral deflection. The differentials in Equation (5.6) are partial since $x$ is also a function of time.

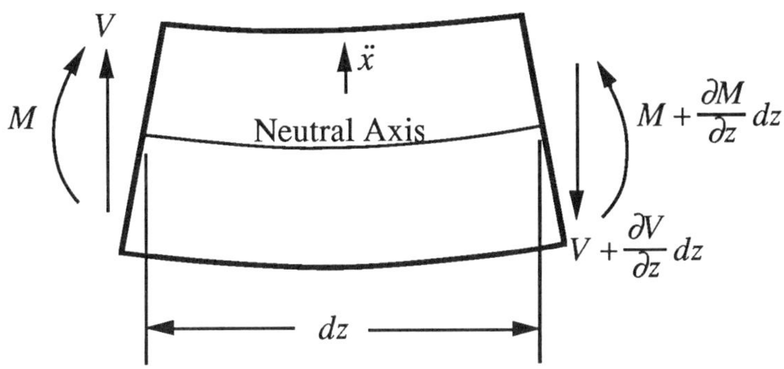


**Figure 5.1** Differential Element of a Beam in Bending

For a vibrating beam, neglecting rotatory inertia, the change in shear across the differential beam element is related to the deflection by

$$\frac{\partial V}{\partial z} = -\rho A \frac{\partial^2 x}{\partial t^2} \tag{5.7}$$

where $V$ is the shear, $\rho$ is the density of the beam material, $A$ is the cross-sectional area of the beam, and $t$ is time. Equations (5.6) and (5.7) may be combined together with the relationship given by

$$V = \frac{\partial M}{\partial z} \tag{5.8}$$

to give the governing equation

$$\frac{\partial^2}{\partial z^2}\left( EI \frac{\partial^2 x}{\partial z^2} \right) = -\rho A \frac{\partial^2 x}{\partial t^2} \tag{5.9}$$

For a uniform beam, $E$, $I$, $\rho$ and $A$ are constants and Equation (5.9) may be reduced to

$$\frac{\partial^4 x}{\partial z^4} = -\frac{\rho A}{EI}\frac{\partial^2 x}{\partial t^2} = -\kappa^2 \frac{\partial^2 x}{\partial t^2} \tag{5.10}$$

where $\kappa^2$ is a positive real value, in this case a constant, which represents the ratio of the mass per unit length of the beam to its flexural rigidity.

Equation (5.10) may be solved by separation of variables where $x$ is taken to be

$$x(z,t) = Z(z)T(t) \tag{5.11}$$

where $Z$ has the same dimensions as $x$ (i.e., length) and $T$ is dimensionless. Substituting Equation (5.11) into Equation (5.10) and rearranging gives

$$\frac{1}{Z}\frac{d^4 Z}{dz^4} = -\frac{\kappa^2}{T}\frac{d^2 T}{dt^2} = \lambda^4 \tag{5.12}$$

where the derivatives are now ordinary and the use of the separation constant, $\lambda^4$, (determined from the boundary conditions) reflects the fact that the two sides of the equation are independent [73]. As will be seen below, $\lambda$, which has the reciprocal dimensions of $x$, becomes part of the arguments of the transcendental solutions to Equation (5.12). Equation (5.12) may be separated into the two equations

$$\frac{d^4 Z}{dz^4} - \lambda^4 Z = 0 \tag{5.13}$$

and

$$\frac{d^2 T}{dt^2} + \frac{\lambda^4}{\kappa^2} T = 0 \tag{5.14}$$

which are solved separately.

The solution to Equation (5.13) is given by

$$Z = A \sin \lambda z + B \cos \lambda z + C \sinh \lambda z + D \cosh \lambda z \tag{5.15}$$

The boundary conditions to be applied to Equation (5.15) are

$$Z(0) = Z(h) = \left.\frac{d^2 z}{dz^2}\right|_{z=0} = \left.\frac{d^2 z}{dz^2}\right|_{z=1} = 0 \tag{5.16}$$

since the bearings act as pin joints or hinges. Substitution of the first and third boundary conditions result in the two equations

$$B + D = 0 \tag{5.17}$$

$$B - D = 0 \tag{5.18}$$

which are satisfied only when $B$ and $D$, are both equal to zero. Applying the other two boundary conditions gives

$$A \sin \lambda l + C \sinh \lambda l = 0 \tag{5.19}$$

$$-A\,\lambda^2 \sin \lambda l \ + C\lambda^2 \sinh \lambda l \ = 0 \tag{5.20}$$

Since sinh $\lambda l$ can never be equal to zero for positive (nonzero) values of $\lambda l$, Equations (5.19) and (5.20) are satisfied only for $C$ equal to zero and sin $\lambda l$ equal to zero. Letting $A$ be equal to zero would result in a trivial solution. Thus, these equations are satisfied for an infinite set of values of $\lambda$ given by

$$\lambda_n = \frac{n\pi}{l} \qquad \text{for } n = 1, 2, 3, \ldots \tag{5.21}$$

where the $\lambda_n$ are referred to as the characteristic values, or eigenvalues, of the rotor system. Substituting back into Equation (5.15) gives

$$Z_n = A_n \sin \frac{n\pi}{l} z = A_n \phi_n(z) \qquad \text{for } n = 1, 2, 3, \ldots \tag{5.22}$$

where the $\phi_n(z)$ are the characteristic functions, or eigenfunctions, of the rotor system. These eigenfunctions are orthogonal and are the same as the normalized mode shapes, $\phi_j(z)$, introduced in Equation (5.3), the first four of which are represented in Figure 5.2.

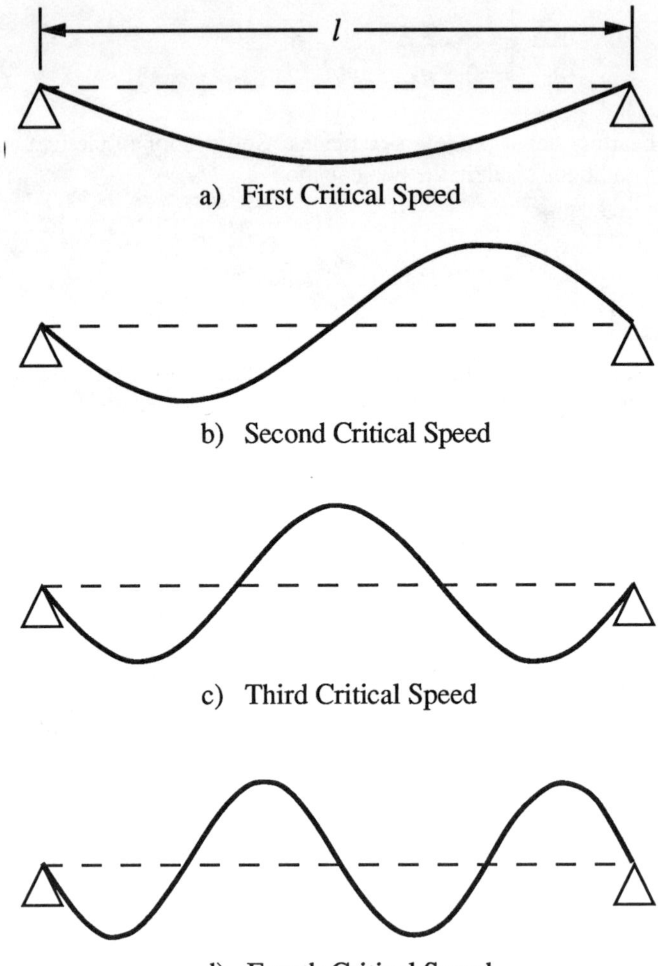


a) First Critical Speed

b) Second Critical Speed

c) Third Critical Speed

d) Fourth Critical Speed

**Figure 5.2** First Four Mode Shapes for a Uniform Flexible Rotor on Rigid Bearings

The total rotor response for any particular speed is generally written in rotating coordinates, without the time varying functions, as a sum of the individual modal response, $Z_n$, as

$$x(z) = \sum_{n=1}^{\infty} A_n \sin \frac{n\pi}{l} z = \sum_{n=1}^{\infty} A_n \phi_n(z) \tag{5.23}$$

where the $A_n$ are coefficients which represent the extent to which each of the modes is excited, and are, in general, functions of the speed of rotation. Each $A_n$ reaches a maximum for a rotational speed equal to the critical speed for the $n^{th}$ mode, $\omega_n$. The values of the critical speeds are determined by solving Equation (5.14) to get

$$T_n = F \sin\left(\frac{\lambda_n^2}{\kappa} t + \theta\right) = F \sin(\omega_n t + \theta) \qquad \text{for } n = 1, 2, 3, \ldots \quad (5.24)$$

where $F$ and $\theta$ are constants which depend upon the initial conditions and $\omega_n$ are the rotor system resonances. These resonances coincide with the critical speeds for an undamped system with negligible gyroscopic effects. Making the substitutions from Equations (5.10) and (5.21), the critical speeds are given by

$$\omega_n = \frac{n^2 \pi^2}{l^2} \sqrt{\frac{EI}{\rho A}} \qquad \text{for } n = 1, 2, 3, \ldots \quad (5.25)$$

For a uniform rotor with different boundary conditions, the solution is similarly straightforward. However, for nonuniform rotors the solutions are much more involved and numerical methods are generally employed.

## Application to Balancing

As discussed above, the orthogonality of these characteristic functions permits them to be used to represent any well-behaved function, using the form of Equation (5.3). Particular functions of interest include rotor mass unbalance and rotor deflection, or vibration amplitude. The rotor mass unbalance may be represented as an eccentricity of the mass centerline, with respect to the bearing centerline. This eccentricity is given as a complex function of $z$ by

$$\hat{a}(z) = a_u(z) + i a_v(z) \quad (5.26)$$

where $a_u(z)$ and $a_v(z)$ are the components of the eccentricity in rotating coordinates. The system of rotating coordinates is represented by $u$ and $v$, as illustrated in Figure 5.3. The superior caret is used to indicate a complex variable. Where all variables in an expression are complex and no ambiguity exists, the superior caret is dropped for convenience.

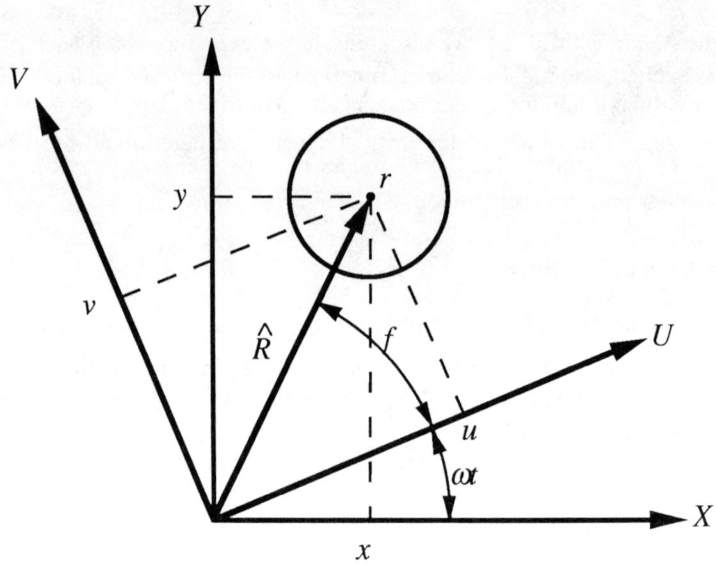


**Figure 5.3** Stationary and Rotating Coordinate Systems for Measurement of Synchronous Rotor Vibration

The complex function $\hat{a}(z)$ can be expanded in a modal, characteristic function, series as

$$\hat{a}(z) = \sum_{j=1}^{\infty} \phi_j(z)\hat{a}_j = \sum_{j=1}^{\infty} \phi_j(z)\,(a_{uj} + ia_{vj}) \tag{5.27}$$

where each of the complex coefficients, $\hat{a}_j$, can, in the form of Equation (5.24), be determined from

$$\hat{a}_j = \frac{1}{Q}\int_0^l A\rho(z)\hat{a}(z)\phi_j(z)\mathrm{d}z \tag{5.28}$$

Similarly, the rotor whirl deformation can be represented in complex notation by

$$\hat{\eta}(z) = \eta_u(z) + i\eta_v(z) \tag{5.29}$$

which can also be expanded as a modal series

$$\hat{\eta}(z) = \sum_{j=1}^{\infty} \phi_j(z)\hat{\eta}_j = \sum_{j=1}^{\infty} \phi_j(z)(\eta_{uj} + i\eta_{vj}) \tag{5.30}$$

Modal balancing procedures are based on the orthogonality relations which require that the response for the $j$th mode, $\phi_j(z)\hat{\eta}_j$, is caused only by the $j$th component of mass eccentricity, $\phi_j(z)\hat{a}_j$. Thus, the individual modal responses are eliminated by removing the corresponding modal components of the mass eccentricity.

The modal response is related to the corresponding modal component of mass eccentricity in the form of the Jeffcott analysis, as given by

$$\hat{\eta}_j = \frac{a_j(\omega/\omega_j)^2 e^{-i\beta_j}}{\sqrt{\left(1 - \omega^2/\omega_j^2\right)^2 + 4(\zeta_j\omega/\omega_j)^2}} \tag{5.31}$$

where $\omega_j$ is the undamped critical speed and $\zeta_j$ is the equivalent viscous damping ratio for the $j$th mode. Equation (5.31) may also be written in the form

$$\hat{\eta}_j = \frac{\hat{a}_j\omega^2}{\omega_j^2 - \omega^2 + 2i\zeta_j\omega_j\omega} \tag{5.32}$$

Note that the total synchronous response for any particular rotational speed is determined from the sum of the modal response for each mode as in Equation (5.30), where the $\hat{\eta}_j$ are complex functions of rotational speed only, and the $\phi_j$ are real functions of $z$ only. In general, the $\omega_j$ and $\zeta_j$ are not known with sufficient accuracy to calculate $\hat{a}_j$ directly from $\hat{\eta}_j$ using Equation (5.31) or (5.32). However, from these equations it can be seen that $\hat{\eta}_j$ is directly proportional to $\hat{a}_j$ and that the proportionality constant, $\hat{\alpha}_j$, is a function of $\omega$ only. This premise, as applied to discrete unbalances, is central to influence coefficient balancing, as well as to the modal balancing procedures, as described below.

It has been shown in reference [18] that the addition of a discrete unbalance with corresponding masses $m_i$, located at axial positions $z_i$, has the effect of changing $\hat{a}_j$ by the amount $\hat{a}_j^*$, given by

$$\hat{a}_j^* = \sum_{i=1}^{n} \frac{m_i \hat{r}_i \phi_j(z_i)}{Q} \tag{5.33}$$

where the $\hat{r}$ are complex and represent the locations of the $m_i$ relative to the center of rotation, in rotating coordinates. The object of modal balancing is to select the $m_i$, $\hat{r}$ and $z_i$ such that $\hat{a}_j^*$ exactly cancels $\hat{a}_j$, resulting in a rotor which is perfectly balanced for the $j^{\text{th}}$ mode. In practice, of course, $\hat{a}_j$ is never exactly cancelled and a rotor is never perfectly balanced. Rather, an acceptable residual $\hat{a}_j$ is generally specified, which is used as a balancing criterion. At the same time, it is necessary to avoid modifying the $\hat{a}_j$ for the previously balanced, usually lower, modes.

## N and (N+2) Plane Modal Balancing

A variety of specific procedures have been used to perform modal balancing. These procedures generally fall into two groups referred to as $N$ plane and ($N$+2) plane modal balancing, typified by references [18] and [71], respectively. For both cases, it is necessary to determine the $\phi_j$ for all resonant modes below the maximum speed and one or two resonant modes above the maximum speed. Usually, this is done analytically, although it is suggested in reference [18] to do this experimentally. It is implied in this reference that by running the rotor at a speed very near to $\omega_j$ the rotor response will approximate $\phi_j$. However, the effect of other, usually higher, unbalanced modes may not be negligible. Thus, mode separation can be a problem. For $N$ plane balancing the set of correction masses for the $N^{\text{th}}$ mode must satisfy

$$\sum_{i=1}^{n} m_i \hat{r}_i \phi_j(z_i) = 0 \qquad \text{for } j = 1, 2, \ldots, N-1 \tag{5.34}$$

in order not to affect the corresponding residual $\hat{a}_j$ for the lower, previously balanced, modes. To eliminate $\hat{a}_N$ (i.e., to compensate for the effect of the $N^{\text{th}}$ mode), these correction masses must also satisfy

$$\sum_{i=1}^{n} m_i \hat{r}_i \phi_N(z_i) = -Q\hat{a}_N \tag{5.35}$$

where $Q$ and $\hat{a}_N$ have been determined from trial mass runs. Generally, the $z_i$ are predetermined and the $m_i$ and $\hat{r}_i$ are combined into a single set of $n$ complex unknowns, each composed of magnitude and angle, where all of the angles are assumed to be identical; defining a planar mass distribution. Thus, Equations (5.34), ($N$–1) real equations, and (5.35), one complex equation, represent $N$ equations with $n$ unknowns; ($n$–1) real and one complex. To obtain a solution for these equations, $n$ must be greater than or equal to $N$. For $N$ plane modal balancing, $n$ is taken equal to $N$ (i.e., to balance the $N^{th}$ mode requires $N$ planes).

The ($N$+2) plane method is very similar except that the rotor is assumed to have been previously balanced as a rigid rotor and two additional constraint equations are added to preserve the rigid rotor balance

$$\sum_{i=1}^{n} m_i \hat{r}_i = 0 \tag{5.36}$$

and

$$\sum_{i=1}^{n} m_i \hat{r}_i z_i = 0 \tag{5.37}$$

At least two additional planes are required to satisfy Equations (5.36) and (5.37) along with Equations (5.34) and (5.35). Thus, ($N$+2) planes are needed to balance the $N^{th}$ mode using this method.

There has been considerable discussion as to whether the use of the two additional planes is justified for practical rotor balancing [3,4]. Clearly, when balancing a rotor that has not been previously balanced as a rigid rotor, it is pointless to satisfy Equations (5.36) and (5.37). Thus, the question becomes whether it is justified to pre-balance a flexible rotor as a rigid rotor. This will depend on two factors: (1) whether the rotor can be mounted on a balancing machine for rigid rotor balancing and then assembled on its own bearings in such a way as to retain the rigid rotor balance (otherwise there is no point in balancing as a rigid rotor); and (2) whether the contributions of the unsatisfied Equations (5.36) and (5.37) have a significant effect on the residual response of the rotor, which is a point of considerable disagreement. Certainly, examples can be, and have been, generated to support either position and, consequently, this discussion has not yet reached a satisfactory conclusion.

The Unified Balancing Approach, the topic of Chapter 7, has been formulated as an $N$ plane procedure. However, the additional constraints of Equations (5.36) and (5.37) can easily be included to result in an ($N$+2) plane procedure. Thus, the Unified Balancing Approach, while fundamentally different from modal balancing, can be configured to acquire the relative advantages of either $N$ plane or ($N$+2) plane modal balancing.

The modal balancing theory, as described above, is based on two important assumptions: (1) the rotor system damping is negligible, and (2) the mode shapes are planar. These assumptions are not always satisfied in practice. As usually applied, the object of modal balancing is to eliminate bearing vibration. Consequently, the residual shaft vibration can be substantial, particularly if the bearings are located on, or near, shaft nodes. Thus, while the calculations required for modal balancing are relatively simple, the procedure requires considerable operator insight and is sensitive to the presence of measurement error, system damping and non-planar mode shapes.

## Advantages and Limitations of Modal Balancing

Specific advantages and disadvantages may be identified with modal balancing, relative to other balancing methods. According to the requirements of a particular balancing situation, these advantages and disadvantages, as enumerated below, take on different relative importance.

The principal advantages of modal balancing are:

1. The number of sensitivity runs required at the highest balancing speed is minimized
2. Good sensitivity at the highest balancing speed can always be achieved
3. Balancing of a specific mode is provided for, while not affecting previously balanced (usually lower) modes
4. It can be an entirely empirical procedure which requires only an understanding of the modal character of the response of a rotor system to unbalance. Although, in present practice, it is most often used in conjunction with analytically determined mode shapes rather than in a solely empirical manner.

The only trial mass runs that are required at the highest balancing speed involve the mass set for the corresponding modal response. Individual trial mass runs are not required at that speed. Since the lower modes are not affected by a modal trial mass set for a higher mode, this mass set can be made large enough to assure good sensitivity at that higher mode. Eliminating the danger of upsetting the condition of previously balanced modes generally simplifies the balancing of higher modes. It is not necessary to have a prior knowledge of the dynamics of the rotor system apart from a basic understanding to assist in the satisfactory selection of locations for vibration sensors and balancing planes. Modal balancing does, of course, assume that the response of the rotor system is linear and that the vibration sensors are sufficiently accurate.

The principal disadvantages of modal balancing are:

1. The assumption of planar modes of vibration inherent in modal balancing may not be valid for systems with substantial damping or bearing cross-coupling effects
2. For most (but not all) applications of modal balancing, accurate prior knowledge of the dynamics of the rotor system is required
3. Effective use requires a high degree of operator insight (i.e., requires a highly skilled operator)
4. Balancing results are generally based on the vibration measured by only one or two sensors for a particular mode, which may not result in a uniformly well-balanced rotor
5. Modal balancing is usually not automated and does not easily lend itself to production applications
6. Often, trimming of lower modes while balancing higher modes is liable to adversely affect the higher modes. However, in theory, this can be avoided with the addition of extra balancing planes

Most applications of modal balancing use analytical procedures for selecting the sets of trial masses for correcting specific modes [69]. Not only does this complicate the balancing procedure, but any inaccuracies in this analytical model, which are inevitable in any analytical representation of real components, tend to reduce the effectiveness of the balancing procedure. It is not necessary, however, to proceed in this way and several proponents of modal balancing employ empirical procedures to determine the modal trial mass sets [18]. However, this empirical procedure requires a substantial amount of operator insight into the modal character of the whirl of a rotor due to unbalance, which is not conducive to the development of efficient production balancing techniques for use by relatively unskilled personnel. Balancing of a rotor to eliminate vibration as measured by a single sensor may not result in a minimum vibration level for the rotor as a whole, particularly when dealing with distorted or non-planar mode shapes. Difficulty may also occur as a result of the presence of vibration from other modes which have not been previously balanced, although methods have been developed for dealing with this problem [15]. Once again, however, operator insight is necessary.

# Chapter 6
# Influence Coefficient Balancing

The assumption that rotor response is proportional to unbalance is basic to virtually all balancing methods for both rigid and flexible rotors. Even for systems where substantial nonlinearity is observed, this assumption can be satisfied by balancing in a stepwise fashion with sufficiently small steps to approximate linear behavior. An additional assumption inherent in most balancing methods, and of particular import to influence coefficient balancing, is that the effect of individual unbalances can be superposed to give the effect of a set of unbalances. This has been generally accepted as a fact for unbalances that are not excessively large. The premise behind influence coefficient balancing is that based on these two assumptions a rotor can be characterized from a set of individual trial mass runs. This characterization can be used to define a combination of these masses which will eliminate, or minimize, the synchronous rotor response due to unbalance. The analytical development and implementation procedures for influence balancing are presented in this chapter. A method for identifying and eliminating non-independent balancing planes is also presented.

## Analytical Development

The influence coefficient balancing procedure, as currently employed, was originally developed by Goodman [60] and subsequently refined by Lund and Tonneson [87] and verified by Tessarzik and others [128, 129]. Using complex notation, the synchronous rotor vibration, as measured in one stationary transverse direction, can be represented as

$$x = |x| \cos(\omega t + \beta) = x_c \cos(\omega t) - x_s \sin(\omega t) \tag{6.1}$$

where

$$x_c = |x| \cos \beta \tag{6.2}$$

$$x_s = |x| \sin \beta \tag{6.3}$$

$|x|$ = vibration amplitude

$\beta$ = leading phase angle

$\omega$ = steady angular speed of shaft rotation

This vibration amplitude may also be represented with the conventional complex notation by

$$\hat{x} = x_c + ix_s \tag{6.4}$$

which is a short form representing

$$x = \text{Real}\left[(x_c + ix_s)\, e^{i\omega t}\right] \tag{6.5}$$

In a similar manner, rotor mass unbalance can be represented in complex form by

$$\hat{u} = u_c + iu_s \tag{6.6}$$

where $u_c$ and $u_s$ are the components of mass unbalance in rotating coordinates. The angle between the stationary coordinate axis along which $x$ is measured and the rotating coordinate axis along which the $u_c$ component is taken is $\omega t$. As virtually all variables in this chapter are complex, the superior carets are dropped for convenience.

## Calculation of Correction Masses

For the balancing procedure, $m$ balancing planes (axial locations) are used for the addition of correction masses. A total of $n$ vibration measurements are taken to determine the response of the rotor. In general, these $n$ measurements consist of data from $n_s$ sensors for each of the $n_\omega$ balancing speeds such that

$$n = n_s n_\omega \tag{6.7}$$

However, in theory, it is not necessary that data be used from the same number of sensors, or even from the same sensors, for each balancing speed. For a linear system, the vibration measurements, represented by the column vector $x_w$ of length $n$, in response to masses placed in each of the $m$ individual planes is given by

$$\mathbf{x}_w = \mathbf{A}\mathbf{w} \tag{6.8}$$

where $\mathbf{w}$ is a column vector of length $m$ representing the masses in these planes and $\mathbf{A}$ is an $n$ by $m$ matrix whose elements, $\alpha_{ij}$, are the influence coefficients relating the rotor response for the specified sensors and speeds to the balancing planes. Throughout this book, bold type is used to indicate multidimensional vector (as distinguished from complex) and matrix quantities; with upper case letters used for matrices and lower case letters used for vectors. The $\alpha_{ij}$ are determined empirically through the application of known trial masses in each of the balancing planes. In the simplest case, a single trial mass is used for each plane, one plane at a time, and

$$\alpha_{ij} = \frac{x_{ij} - x_{i0}}{T_j} \tag{6.9}$$

where $x_{i0}$ is the $i^{\text{th}}$ vibration reading with no trial masses installed, $x_{ij}$ is the $i^{\text{th}}$ vibration reading with a trial mass installed in the $j^{\text{th}}$ balancing plane, and $T_j$ is a complex value representing the amplitude and angular location, in rotating coordinates, of this trial mass. Thus, ($m$+1) test runs, including the run with no trial masses, are required to provide sufficient data to calculate all $\alpha_{ij}$, where n measurements are taken for each test run.

Once the influence coefficients have been determined, the appropriate correction masses, $\mathbf{w}_m$, can be calculated. The vibration which remains after the correction masses are installed, referred to as residual vibration, is given by

$$\mathbf{x} = \mathbf{x}_0 + \mathbf{x}_m \tag{6.10}$$

where $\mathbf{x}_0$ is the vector of complex vibration amplitudes measured prior to the installation of the correction masses, and $\mathbf{x}_m$ is that due to the correction mass set which, from Equation (6.8), is related to the correction masses by

$$\mathbf{x}_m = \mathbf{A}\mathbf{w}_m \tag{6.11}$$

Substituting Equation (6.11) into Equation (6.10) results in

$$\mathbf{x} = \mathbf{x}_0 + \mathbf{A}\mathbf{w}_m \tag{6.12}$$

The object of the balancing procedure is to minimize $\mathbf{x}$.

For the case in which the number of measurements is equal to the number of balancing planes ($n = m$), Equation (6.12) may be solved directly for all elements of $\mathbf{x}$ equal to zero to give

$$\mathbf{w}_m = -\mathbf{A}^{-1}\mathbf{x}_0 \tag{6.13}$$

where $\mathbf{A}^{-1}$ represents the matrix inverse of the square matrix $\mathbf{A}$, assuming $\mathbf{A}$ is not singular,

However, for the case in which the number of measurements exceeds the number of balancing planes ($n > m$), $\mathbf{x}$ cannot generally be reduced to zero. Instead, following the procedure of Lund and Tonneson [87], the object is to minimize the sum of squares, $s$, of the elements of $\mathbf{x}$ where

$$s = \sum_{i=1}^{n} |x_i|^2 = \sum_{i=1}^{n} \bar{x}_i x_i = \bar{\mathbf{x}}^T \mathbf{x} \tag{6.14}$$

where the superscript bar represents the complex conjugate and the superscript $T$ represents the transpose of a vector or matrix. Together the bar and $T$ represent the complex conjugate transpose of a vector or matrix. Also, the $x_i$ represent the elements of $\mathbf{x}$, and $s$ is always positive and real. In standard linear analysis fashion, $s$ is minimized by the use of

$$\frac{\partial s}{\partial \bar{w}_{mj}} = 0 = \sum_{i=1}^{n} x_i \frac{\partial \bar{x}_i}{\partial \bar{w}_{mj}} = \sum_{i=1}^{n} x_i \bar{\alpha}_{ij} \qquad \text{for } j = 1,2,\ldots,m \tag{6.15}$$

where the $w_{mj}$ are the elements of $\mathbf{w}_m$. Equation (6.12) may be rewritten for the individual complex elements of the vectors as

$$x_i = x_{0i} + \sum_{j=1}^{m} \alpha_{ij} w_{mj} \qquad \text{for } i = 1,2,\ldots,n \tag{6.16}$$

where the $x_{0i}$ are the elements of $\mathbf{x}_0$. Substituting Equation (6.16) into Equation (6.15) gives

$$\sum_{i=1}^{n} \bar{\alpha}_{ij} \left( x_{0i} + \sum_{h=1}^{m} \alpha_{ih} w_{mh} \right) \qquad \text{for } j = 1,2,\ldots,m \tag{6.17}$$

Equation (6.17) may be rearranged and rewritten in matrix form yielding

$$\bar{\mathbf{A}}^T \mathbf{A} \mathbf{w}_m + \bar{\mathbf{A}}^T \mathbf{x}_0 = 0 \qquad (6.18)$$

Since $\bar{\mathbf{A}}^T \mathbf{A}$ is square, Equation (6.18) may be solved directly for $\mathbf{w}_m$, assuming $\bar{\mathbf{A}}^T \mathbf{A}$ is not singular, resulting in

$$\mathbf{w}_m = -\left(\bar{\mathbf{A}}^T \mathbf{A}\right)^{-1} \bar{\mathbf{A}}^T \mathbf{x}_0 \qquad (6.19)$$

This is the set of correction masses which minimizes the sum of squares, or similarly the root mean square, of the residual vibration. The expected residual vibration can be calculated by substituting Equation (6.19), or the numerical result of that equation, into Equation (6.12). While this least squares procedure provides for a minimization of the sum of the squares of the residual vibration amplitudes, it does not ensure that none of the individual residual vibration measurements remains large. In fact, it is not unusual for one or two of the vibration readings to remain quite large, while the remainder are reduced substantially. If this situation is undesirable and it is preferred to minimize the maximum residual amplitude (i.e., the largest element of $\mathbf{x}$), an iterative procedure may be used, which was developed by Goodman [60]. In general, this procedure tends to equalize the elements of $\mathbf{x}$ at the expense of the root mean square average of these elements.

To perform this procedure, Equation (6.14) is rewritten as

$$s_k = \sum_{i=1}^{n} \varepsilon_{ik} \, | x_i |_k^2 = \text{minimum} \qquad \text{for } k = 0,1,2,\ldots \qquad (6.20)$$

where $k$ is the iteration number and $\varepsilon_{ik}$ is a weighting factor which is given by

$$\varepsilon_{ik} = \prod_{p=0}^{k-1} \frac{| x_i |_p}{(\text{rms})_p} = \frac{| x_i |_{k-1}}{(\text{rms})_{k-1}} \varepsilon_{i,k-1} \qquad (6.21)$$

where $\prod$ denotes a multiplication series, $(\text{rms})_p$ is the real-valued, root mean square average of the elements of the vector $\mathbf{x}$ from the $p^{\text{th}}$ iteration of the correction mass calculation, and $\varepsilon_{ik}$ has a positive real value. Thus

$$(\text{rms})_p = \sqrt{\frac{\sum_{i=1}^{n} |x_i|_p^2}{n}} = \sqrt{\frac{1}{n}\left(\bar{\mathbf{x}}^T\mathbf{x}\right)_p} \tag{6.22}$$

The first iteration ($k = 0$) is identical to the least squares calculation described above, Equation (6.19), which is obtained for

$$\varepsilon_{i0} = 1 \qquad \text{for } i = 1,2,\ldots,n \tag{6.23}$$

A result analogous to Equation (6.19) including the effect of the weighting factors is found in a similar manner as above. The final result is

$$\mathbf{w}_{mk} = -\left(\bar{\mathbf{A}}^T\mathbf{E}_k\mathbf{A}\right)^{-1}\bar{\mathbf{A}}^T\mathbf{E}_k\mathbf{x}_0 \tag{6.24}$$

where $\mathbf{w}_{mk}$ is the $k^{\text{th}}$ iteration of $\mathbf{w}_m$, and $\mathbf{E}_k$ is an $n$ by $n$ diagonal matrix in the form

$$\mathbf{E}_k = \begin{bmatrix} \varepsilon_{1k} & 0 & 0 & 0 \\ 0 & \varepsilon_{2k} & 0 & 0 \\ 0 & 0 & \cdots & \cdots \\ 0 & 0 & \cdots & \varepsilon_{nk} \end{bmatrix} \tag{6.25}$$

A convergence limit can be specified for the iteration procedure such that when

$$\frac{\left|\,|x_i|_k - |x_i|_{k-1}\right|_{\max}}{(\text{rms})_k} \leq \text{convergence limit} \tag{6.26}$$

the iteration procedure is considered to have converged.

It should be noted that the results of this least squares procedure, with or without the weighting factors, are dependent on the scaling of the vibration readings. Thus, to obtain meaningful results, it is generally advisable that the vibrations be measured in the same units and have the same order of magnitude. This can be accomplished by manipulation of calibration factors for the various sensors. By the same token, if it is desirable for the residual vibration at one or more particular sensors or balancing speeds to be less than that for the remaining sensors or speeds, a magnification factor can be applied to the corresponding elements of the weighting matrix before the application of Equation (6.24). In this way, a considerable amount of flexibility can be introduced into the procedure for calculating correction masses. Such a magnification factor is applied for a particular vibration reading by multiplying the element in the appropriate row of the matrix $\mathbf{E}_k$ by this factor. This is equivalent to multiplying $\mathbf{E}_k$ by the diagonal matrix, $\mathbf{F}$, given by

$$\mathbf{F} = \begin{bmatrix} f_1 & 0 & 0 & 0 \\ 0 & f_2 & 0 & 0 \\ 0 & 0 & \dots & \dots \\ 0 & 0 & \dots & f_n \end{bmatrix} \tag{6.27}$$

where $f_i$ is the magnification factor for the $i^{\text{th}}$ row of $\mathbf{E}_k$.

## Calculation of Influence Coefficients

The influence coefficients may be most directly calculated using Equation (6.9), once the appropriate vibration data has been collected. However, there are certain errors inherent in the measurement of real vibration data due to sensor and instrumentation inaccuracies and rotor system nonlinearities. Consequently, Lund and Tonneson [87] developed a procedure by which a second trial mass is used in the same plane and the two sets of trial mass data are used to optimize the influence coefficients and reduce the effect of these inherent measurement errors. The analytical development of this procedure is replicated below.

If the measurement errors for $x_{ij}$ and $x_{i0}$ are taken to be $\Delta x_{ij}$ and $\Delta x_{i0}$, respectively, Equation (6.9) is modified to give

$$\alpha_{ij} = \frac{(x_{ij} + \Delta x_{ij}) - (x_{i0} + \Delta x_{i0})}{T_j} \tag{6.28}$$

where $\alpha_{ij}$ is now the *true* influence coefficient. A second trial mass, $T_j'$, which is different than $T_j$, is installed in the same balancing plane and the corresponding vibration, $x_{ij}'$, is measured. The corresponding influence coefficient $a_{ij}'$, is calculated from

$$\alpha_{ij}' = \frac{(x_{ij}' + \Delta x_{ij}') - (x_{i0} + \Delta x_{i0})}{T_j'} \tag{6.29}$$

where $\Delta x'_{ij}$ is the measurement error corresponding to $x'_{ij}$. Since $\alpha_{ij}$ and $\alpha'_{ij}$ are the same, the right sides of Equations (6.28) and (6.29) can be equated and rearranged to yield

$$\Delta x_{i0} = b_{ij} + c_j \Delta x_{ij} + d_j \Delta x'_{ij} \tag{6.30}$$

where

$$b_{ij} = c_j(x_{ij} - x_{i0}) + d_j(x'_{ij} - x_{i0}) \tag{6.31}$$

$$c_j = \frac{T'_j}{T_j - T'_j} \tag{6.32}$$

$$d_j = \frac{T_j}{T_j - T'_j} \tag{6.33}$$

and all values are complex.

The errors may then be determined by minimizing $s_{ij}$ where

$$s_{ij} = e_{ij} \mid \Delta x_{i0} \mid^2 + \mid \Delta x_{ij} \mid^2 + \mid \Delta x'_{ij} \mid^2$$

$$= e_{ij} \Delta x_{i0} \overline{\Delta x_{i0}} + \Delta x_{ij} \overline{\Delta x_{ij}} + \Delta x'_{ij} \overline{\Delta x'_{ij}} \tag{6.34}$$

and $s_{ij}$ and $e_{ij}$ are real values. Since $\Delta x_{i0}$ is used for $s_{ij}$ for all m values of $j$, the weighting factor $e_{ij}$ is used to fractionalize the $\Delta x_{i0}$ term among the corresponding m equations. Consequently, for any single balancing sensor, the $e_{ij}$ must sum to one. That is,

$$\sum_{j=1}^{m} e_{ij} = 1 \qquad \text{for } i = 1,2,\ldots,n \tag{6.35}$$

Then, the $s_{ij}$ are minimized by taking

$$\frac{\partial s_{ij}}{\partial(\overline{\Delta x_{ij}})} = 0 = e_{ij} \Delta x_{i0} \frac{\partial(\overline{\Delta x_{i0}})}{\partial(\overline{\Delta x_{ij}})} + \Delta x_{ij} \tag{6.36}$$

$$\frac{\partial s_{ij}}{\partial(\overline{\Delta x'_{ij}})} = 0 = e_{ij} \Delta x_{i0} \frac{\partial(\overline{\Delta x_{i0}})}{\partial(\overline{\Delta x'_{ij}})} + \Delta x'_{ij} \tag{6.37}$$

The partial derivatives on the right sides of Equations (6.36) and (6.37) may be evaluated from the complex conjugate of Equation (6.30) to yield

$$\frac{\partial(\overline{\Delta x_{i0}})}{\partial(\overline{\Delta x_{ij}})} = \bar{c}_j \tag{6.38}$$

$$\frac{\partial(\overline{\Delta x_{i0}})}{\partial(\overline{\Delta x'_{ij}})} = \bar{d}_j \tag{6.39}$$

Substituting Equations (6.38) and (6.39) into Equations (6.36) and (6.37), respectively, gives

$$e_{ij}\bar{c}_j \Delta x_{i0} + \Delta x_{ij} = 0 \tag{6.40}$$

$$e_{ij}\bar{d}_j \Delta x_{i0} + \Delta x'_{ij} = 0 \tag{6.41}$$

Equations (6.40) and (6.41) may be combined to give

$$\bar{d}_j \Delta x_{ij} = \bar{c}_j \Delta x'_{ij} \tag{6.42}$$

Then, Equation (6.42) may be solved for either $\Delta x_{ij}$ or $\Delta x'_{ij}$ and substituted into Equation (6.30) to yield the alternative forms

$$\Delta x_{i0} = b_{ij} + \frac{|c_j|^2 + |d_j|^2}{\bar{c}_j} \Delta x_{ij} \tag{6.43}$$

$$\Delta x_{i0} = b_{ij} + \frac{|c_j|^2 + |d_j|^2}{\bar{d}_j} \Delta x'_{ij} \tag{6.44}$$

Solving Equations (6.43) and (6.44) for $\Delta x_{ij}$ and $\Delta x'_{ij}$, respectively, yields

$$\Delta x_{ij} = \frac{\bar{c}_j}{|c_j|^2 + |d_j|^2} (\Delta x_{i0} - b_{ij}) \tag{6.45}$$

$$\Delta x'_{ij} = \frac{\bar{d}_j}{|c_j|^2 + |d_j|^2} (\Delta x_{i0} - b_{ij}) \tag{6.46}$$

Substituting either Equation (6.45) into Equation (6.40), or Equation (6.46) into Equation (6.41), and solving for $e_{ij}$ yields

$$e_{ij} = \frac{b_{ij} - \Delta x_{i0}}{\left(|c_j|^2 + |d_j|^2\right)\Delta x_{i0}} \tag{6.47}$$

By applying the constraint of Equation (6.35) to Equation (6.47), $e_{ij}$ can be eliminated. Then solving for $\Delta x_{i0}$ gives

$$\Delta x_{i0} = \frac{\displaystyle\sum_{j=1}^{m} \frac{b_{ij}}{|c_j|^2 + |d_j|^2}}{1 + \displaystyle\sum_{j=1}^{m} \frac{1}{|c_j|^2 + |d_j|^2}} \tag{6.48}$$

Substituting Equations (6.31) through (6.33) into Equation (6.48) results in Equation (6.49) for $\Delta x_{i0}$ as a function of known quantities only.

$$\Delta x_{i0} = \frac{\displaystyle\sum_{j=1}^{m} \frac{\overline{T}_j - \overline{T}'_j}{|T_j|^2 + |T'_j|^2}\left[T_j\left(x'_{ij} - x_{i0}\right) - T'_j\left(x_{ij} - x_{i0}\right)\right]}{1 + \displaystyle\sum_{j=1}^{m} \frac{|T_j - T'_j|^2}{|T_j|^2 + |T'_j|^2}} \tag{6.49}$$

Similarly, equations for $\Delta x_{ij}$ and $\Delta x'_{ij}$ can be formed from Equations (6.43) and (6.44) to give

$$\Delta x_{ij} = \frac{\overline{T}'_j}{|T_j|^2 + |T'_j|^2}\left[T_j\left(x'_{ij} - x_{i0} - \Delta x_{i0}\right) - T'_j\left(x_{ij} - x_{i0} - \Delta x_{i0}\right)\right] \tag{6.50}$$

$$\Delta x'_{ij} = \frac{-\overline{T}_j}{|T_j|^2 + |T'_j|^2}\left[T_j(x'_{ij}-x_{i0}-\Delta x_{i0}) - T'_j(x_{ij}-x_{i0}-\Delta x_{i0})\right] \qquad (6.51)$$

Then, by substituting either Equation (6.50) into Equation (6.28), or Equation (6.51) into Equation (6.29), the relation for the adjusted value of the influence coefficient is obtained in the form

$$\alpha_{ij} = \frac{1}{|T_j|^2 + |T'_j|^2}\left[\overline{T}_j(x_{ij} - x_{i0} - \Delta x_{i0}) + \overline{T}'_j(x'_{ij} - x_{i0} - \Delta x_{i0})\right] \qquad (6.52)$$

Also, the corresponding element of the uncorrected rotor data vector, $x_{i0}$, is modified by adding $\Delta x_{i0}$ before performing any correction mass calculations.

Often in practice, the second trial mass, $T'_j$, is equal in magnitude and 180 degrees out of phase with the first trial mass, $T_j$, for the same balancing plane (i.e., $T'_j = -T_j$). In this case, Equation (6.52) reduces to

$$\alpha_{ij} = \frac{x_{ij} - x'_{ij}}{2T_j} \qquad (6.53)$$

which is independent of $\Delta x_{i0}$. However, even for this special case, $x_{i0}$ is modified by adding the correction $\Delta x_{i0}$, and thus the calculated correction masses are affected by $\Delta x_{i0}$.

This procedure permits both the optimization of influence coefficients and the evaluation of measurement errors. This is useful not only for improving the balancing calculation, but also for identifying erroneous data. If two trial mass runs per balancing plane are used, but, for some reason, it is not desired to optimize the influence coefficients or evaluate the measurement errors, the $\Delta x_{i0}$ are assumed equal to zero and Equation (6.52) is reduced to give the equation

$$\alpha_{ij} = \frac{\overline{T}_j(x_{ij} - x_{i0}) + \overline{T}'_j(x'_{ij} - x_{i0})}{|T_j|^2 + |T'_j|^2} \qquad (6.54)$$

for the calculation of influence coefficients. In addition, the $x_{i0}$ are not modified prior to the correction mass calculations.

A procedure has been developed by Larsson [77] for calculating influence coefficients from any number of balancing runs in which one or more trial and/or correction masses has been added to the rotor, using a statistical regression algorithm. This procedure allows the values of the influence coefficients to be continually refined. However, the uncorrected rotor data are not modified to account for measurement error as with Lund's method [87].

As discussed above, the influence coefficient procedure is based on the assumptions of rotor linearity and superposition of rotor unbalance. An additional assumption inherent in the matrix manipulations described above is that the balancing planes are independent. That is, that the influence coefficient matrix, **A**, is of rank $m$ (i.e., **A** is well conditioned). This assumption is often, unknowingly, violated. The result can be the calculation of large correction masses in the non-independent planes, which essentially cancel each other and cannot practically be installed. This potential problem is discussed in more detail below.

## Procedure for Influence Coefficient Balancing

When applying the influence coefficient procedure, some prior knowledge of the dynamics of the rotor system is useful when choosing the location and number of balancing planes and sensors, choosing the balancing speeds, and estimating safe vibration limits. The balancing procedure is begun by taking uncorrected rotor data. Then, either one or two trial masses are installed in each balancing plane, one at a time, and trial mass data is taken. In general, the trial masses are removed. However, if the software is designed to accommodate it, the trial masses may be left in. The influence coefficient matrix, one or more sets of correction masses and the corresponding expected residual vibration are then calculated, as outlined above. Except for fairly simple cases, a computer is required to perform these calculations, although this need not be a large computer [28]. A set of correction masses is chosen and installed and check balance data are taken. In many cases, this is the end of the balancing procedure. However, if the check balance data is not satisfactory, or if vibration must be controlled at a higher speed, it is necessary to perform another balancing run and the procedure described above is repeated.

Under certain circumstances, existing influence coefficients may be used for balancing a rotor, eliminating the requirement for trial mass runs. In particular, trim balancing a rotor which is relatively linear at a speed, or speeds, at which it has been previously balanced can often be accomplished using existing influence coefficients. Such trim balancing is sometimes necessary during the initial balancing procedure for a rotor (i.e., balancing in steps) and may also be used during maintenance of the same rotor system, if its dynamics have not changed substantially. Existing influence coefficients can also often be used for an identical rotor, if the dynamics are very similar. This is sometimes referred to as one-shot balancing procedure. This procedure may be optimized by extracting influence coefficients from both trial and correction mass runs for many identical rotors and then using statistical methods for averaging these influence coefficients, as mentioned above [77].

It is sometimes required to change balancing speeds between balancing runs, particularly when it becomes necessary to take data at a speed nearer to a critical speed in order to increase the sensitivity of the rotor to the residual unbalance. Since the influence coefficients change with speed, it is necessary to use a different set of influence coefficients, and, unless existing influence coefficients are available for the new balancing speed, a full set of trial mass runs are required. This can be very time-consuming, particularly if a large number of balancing planes are used.

## Advantages and Limitations of Influence Coefficient Balancing

As with modal balancing, specific advantages and disadvantages may be identified with influence coefficient balancing, a number of which are enumerated below.

The principal advantages of influence coefficient balancing are:

1. It is an entirely empirical procedure which requires minimal prior knowledge of the dynamics of the rotor system
2. Full sensitivity information, if available, allows convenient balancing of any combination of critical speeds (modes), provided the rotor can be safely operated at the speeds for the measurement of the unbalanced whirl
3. It is readily computerized and automated
4. It provides for least-squares minimization of data from any number of vibration sensors
5. Data manipulation techniques have been developed which compensate for measurement errors
6. It does not require a high degree of operator insight

Other than a minimum knowledge of the rotor system, which is required for specifying the locations of the vibration sensors and balancing planes for the application of correction masses, influence coefficient balancing is an entirely empirical procedure whose effectiveness is theoretically limited only by the range of linearity of the rotor system and the resolution of the vibration sensors. Influence coefficient balancing provides for the simultaneous balancing of more than one mode, provided that the necessary sensitivity information (i.e. set of influence coefficients) is available. This can result in a significant reduction in the required number of balancing runs, but does require measurement of the rotor vibration at speeds near the relevant critical speeds. It may not be possible to safely run through one of these critical speeds to make measurements at higher speeds.

As to the advantages of this method, the empirical nature of this procedure and the modest skills required of the operator, along with the form of the required calculations, make this procedure ideally suited for computerization and automation. Least-squares minimization of vibration data is used to permit the application of balancing criteria based on reducing the vibration at a number of points in the rotor system for any number of modes. This feature of influence coefficient balancing can often be used to compensate for distorted or non-planar mode shapes and the effect of other unbalanced modes. Additional data manipulation techniques are available, which are designed to compensate for measurement errors inherent in any realistic application of rotor balancing.

Turning to the principle disadvantages of influence coefficient balancing:

1. A significant number of runs is required to obtain rotor sensitivity data at the highest balancing speed
2. When the lowest modes are the most lightly damped, it is often difficult to get accurate sensitivity data near the higher modes
3. Data must be taken at previously balanced modes in order not to adversely affect the condition of these modes when balancing other modes
4. The least-squares procedure often results in the deterioration of previously balanced modes, unless these modes are heavily weighted, in which case the other modes may receive insufficient emphasis
5. The inadvertent use of non-independent balancing planes often results in the calculation of impractical, and generally inappropriate, results, such as large mutually counteracting correction masses

Since individual trial mass runs must be made at the highest balancing speed, as at all balancing speeds, when using influence coefficient balancing, the total number of runs required at this highest speed are substantially more than for modal balancing. In addition, the sizes of the individual trial masses to be used at the higher balancing speeds may be limited by the response of lightly damped lower modes to these trial masses. The initial unbalance present for the higher modes may make it impossible to acquire sensitivity data, using a restricted trial mass, at a speed close enough to a higher critical speed to produce a significant effect for this trial mass. That is, if the largest possible trial mass is small compared to the initial unbalance at the higher mode, the resolution of the sensitivity data acquired for this trial mass may be substantially reduced. When balancing a particular mode, after having previously balanced a number of other modes, a large quantity of data may be required which can ultimately affect the efficiency of the balancing procedure in terms of both convenience and accuracy.

The inadvertent use of non-independent balancing planes can result in singular, or near singular, influence coefficient matrices. In the case of a singular matrix, a computer program makes this condition known to the operator who can then change the balancing configuration. Due to noise in the vibration data signals and the effect of other unbalanced modes, it is virtually impossible for the influence coefficient matrix to be exactly singular. However, this matrix can be near singular (i.e., having a very small determinant), in which case the data would not be rejected by a computer program, but would often result in the calculation of fairly large correction masses which essentially cancel each other, insofar as their effect on the mode being balanced is concerned. Darlow [27] discussed this very serious problem in much more detail and described a balancing plane optimization procedure which can be used to detect the existence of an ill conditioned (i.e. nearly singular) influence coefficient matrix and eliminate the redundant balancing planes. This balancing plane optimization procedure is described below.

## Elimination of Non-Independent Balancing Planes

As discussed above, the influence coefficient balancing calculations are prone to certain numerical difficulties when non-independent balancing planes are inadvertently used, resulting in impractical, artificially large, mutually counteracting correction masses being calculated. This occurs as a result of the influence coefficient matrix being ill conditioned. That is, at least one of the columns of the matrix is close to being linearly dependent on the remaining columns.

It is possible to anticipate and avoid this problem by evaluating the independence of the columns of the influence coefficient matrix. If this matrix is found to be ill conditioned, it is necessary to identify and eliminate the *least independent* of the balancing planes. This can be done by means of a Gram-Schmidt orthogonalization procedure for optimizing balancing plane selection. The theoretical development of this procedure, from Darlow [27], is presented below along with the step-by-step algorithm required for its implementation.

## Analytical Development of the Orthogonalization Procedure

Recall that the influence coefficient balancing procedure involves the solution of the complex valued matrix equation

$$\mathbf{A}\mathbf{w} = -\mathbf{x}_{\mathrm{w}} \tag{6.55}$$

where $\mathbf{A}$ is the influence coefficient matrix, $\mathbf{x}$ is a column vector containing the unbalanced rotor vibration data and $\mathbf{w}$ is the unknown column vector of the correction masses. As above, bold type is used to indicate multidimensional vector or matrix quantities. Virtually all variables in this development are complex. Thus, the superior caret is not used.

The influence coefficient matrix may be either square or rectangular. If this matrix is square, the solution of Equation (6.55), whether by inversion or decomposition, is straightforward. In the case of a rectangular matrix, a least-squares procedure is used to calculate the correction masses, as described above.

When correction planes are chosen that are not all clearly independent, certain problems can occur with the solution. Mathematically, as the influence coefficient matrix becomes singular, the solution vector becomes infinite. To demonstrate what can occur, the following presentations of the matrix **A** and the vector **x** are proposed:

$$\mathbf{A} = \begin{bmatrix} a & ca \\ b & cb + \varepsilon_1 \end{bmatrix} \quad \text{and} \quad \mathbf{x} = \begin{bmatrix} da \\ db + \varepsilon_2 \end{bmatrix} \tag{6.56}$$

for a simple case for which **A** is square and of rank 2 (if $\varepsilon_1 \neq 0$). The value of $\varepsilon_1$ determines the extent of ill conditioning of the matrix **A**. Substituting Equations (6.56) into Equation (6.55) and solving for **w** yields

$$\mathbf{w} = \begin{bmatrix} c\,\dfrac{\varepsilon_2}{\varepsilon_1} - d \\ -\dfrac{\varepsilon_2}{\varepsilon_1} \end{bmatrix} \tag{6.57}$$

The term $(-d)$ in Equation (6.57) represents the proper correction mass based on the first column of **A** and the first element of **x**, as illustrated in Figure 6.1. The remaining terms in Equations (6.57) represent the effect of the second column of **A** and the second element of **x** on the correction mass calculation. For small values of $\varepsilon_1$, resulting in a nearly singular matrix **A**, these terms may become unreasonably large, as illustrated in Figure 6.2. Clearly, the magnitude of these terms is also dependent on the value of $\varepsilon_2$. However, experimental results have indicated that the near singularity of the influence coefficient matrix, rather than the form of the vector **x,** is more often the determining factor. Thus, if the nearly dependent columns of the influence coefficient matrix are eliminated as in Figure 6.1, or constrained, the resulting correction mass calculation is almost certain to give reasonable results. Consequently, the first and most important step in this optimization procedure is the elimination of the nearly dependent columns of the influence coefficient matrix. The elimination, or constraint, of the corresponding planes will then ensure a well-behaved solution to Equation (6.55) regardless of the form of the vector **x**.

A Gram-Schmidt orthogonalization procedure [99] can be used to eliminate the nearly dependent columns from the influence coefficient matrix. In practice, the optimum balancing planes are those that are not only independent but result in the smallest possible correction masses. That is, if two balancing planes are specified and are found not to be independent, the better balancing plane to use would be the one which would require the smaller correction mass. In general, this would be indicated by the column of the matrix with the largest Euclidean norm. For this reason, the first step in this procedure is to express the influence coefficient matrix as a group of column vectors and to rearrange these vectors in order of decreasing Euclidean norm. These vectors are represented by $\mathbf{u}_1$, $\mathbf{u}_2, \ldots, \mathbf{u}_m$, where $\mathbf{u}_1$ has the largest Euclidean norm and $m$ balancing planes are initially specified. The Euclidean norm of a vector is equal to the square root of the sum of squares of the magnitudes of the elements of that vector.

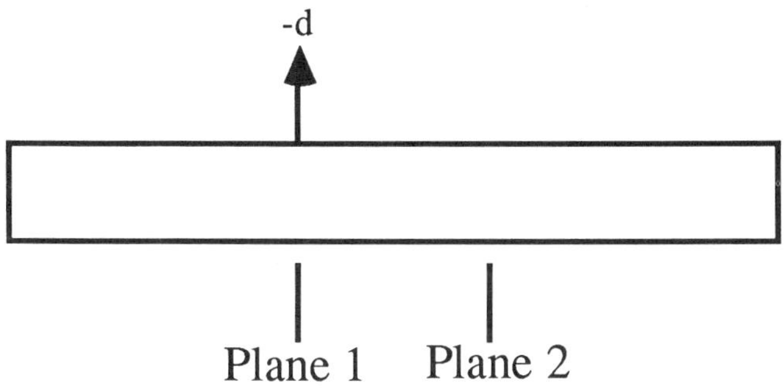


**Figure 6.1** Illustration of Single Plane Correction

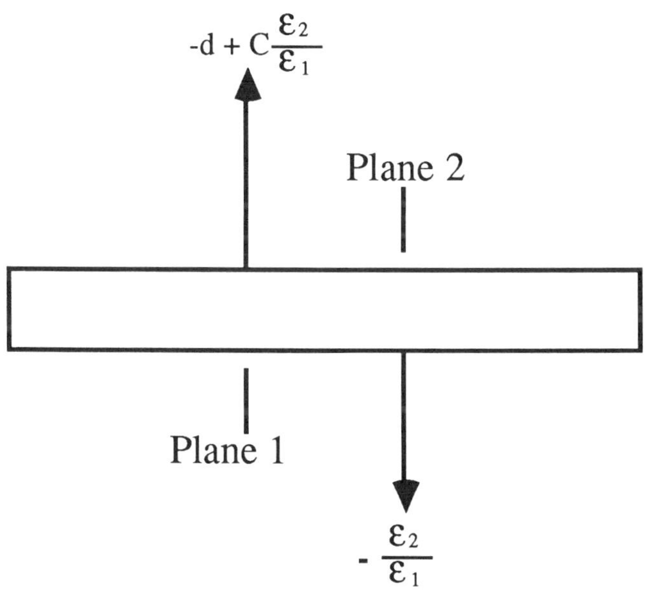


**Figure 6.2** Illustration of Two Plane Correction with Non-Independent Balancing Planes

A basis, or partial basis, is then found for this group of vectors and those vectors that are found not to be *clearly* independent of the previously determined basis vectors are discarded or constrained. A *significance* criteria is used to evaluate independence of the vectors, as described below. The first basis vector is found by normalizing $\mathbf{u}_1$ by

$$\mathbf{e}_1 = \frac{\mathbf{u}_1}{||\mathbf{u}_1||} \tag{6.58}$$

where $\mathbf{e}_i$ represents the $i^{th}$ basis vector and $||\mathbf{u}_i||$ represents the Euclidean norm of $\mathbf{u}_i$. The orthogonal portion of $\mathbf{u}_2$ is given by

$$\mathbf{v}_2 = \mathbf{u}_2 - (\,\bar{\mathbf{e}}_1^T\,\mathbf{u}_2)\,\mathbf{e}_1 \tag{6.59}$$

where the superscript bar represents the complex conjugate, and the superscript $T$ represents the transpose of a vector or matrix. The *significance factor* for $\mathbf{u}_2$ is given by

$$S_{f2} = \frac{||\mathbf{v}_2||}{||\mathbf{u}_2||} \tag{6.60}$$

This significance factor must be, by definition, real-valued, non-negative and no larger than one, since the norm of $\mathbf{v}_2$ can be no larger than the norm of $\mathbf{u}_2$. It is a measure of the independence of $\mathbf{u}_2$ from $\mathbf{e}_1$ (or $\mathbf{u}_1$). A real-valued *independence criterion*, $\delta$, must be chosen as a minimum acceptable value for $S_{f2}$. The smaller the value of $\delta$, the more nearly singular is the influence coefficient matrix. Darlow [27] suggested an appropriate value of $\delta$ to be about 0.2. If $S_{f2}$ is less than $\delta$, then $\mathbf{u}_2$ is to be eliminated or constrained, and $\mathbf{u}_3$ substituted for $\mathbf{u}_2$ in Equations (6.59) and (6.60). However, if $S_{f2}$ is not less than $\delta$, the second basis vector is given by

$$\mathbf{e}_2 = \frac{\mathbf{v}_2}{||\mathbf{v}_2||} \tag{6.61}$$

The orthogonal portion of $\mathbf{u}_3$ is then computed from

$$\mathbf{v}_3 = \mathbf{u}_3 - (\,\bar{\mathbf{e}}_1^T\,\mathbf{u}_3)\,\mathbf{e}_1 - (\,\bar{\mathbf{e}}_2^T\,\mathbf{u}_3)\,\mathbf{e}_2 \tag{6.62}$$

and the corresponding significance factor from

$$S_{f3} = \frac{||\mathbf{v}_3||}{||\mathbf{u}_3||} \tag{6.63}$$

Then $S_{f3}$ is compared to $\delta$ and either $\mathbf{u}_3$ is eliminated or constrained (if $S_{f3} < \delta$) or $\mathbf{e}_3$ is computed from

$$\mathbf{e}_3 = \frac{\mathbf{v}_3}{||\mathbf{v}_3||} \tag{6.64}$$

(if $S_{f3} \geq \delta$). This procedure is repeated for each $\mathbf{u}_i$.

To summarize this procedure, first $\mathbf{e}_1$ is calculated from Equation (6.58). Clearly, the vector with the largest Euclidean norm is always retained since it is used to form the first basis vector. Then, for each $\mathbf{u}_i$, the following sequence of calculations is performed:

1. The orthogonal portion of $\mathbf{u}_i$ is computed from

$$\mathbf{v}_i = \mathbf{u}_i - \sum_{j=1}^{k} (\overline{\mathbf{e}}_j^T \mathbf{u}_i)\, \mathbf{e}_j \tag{6.65}$$

where $k$ basis vectors have thus far been determined ($k \leq i - 1$).

2. The significance factor for $\mathbf{u}_i$ is computed from

$$S_{fi} = \frac{||\mathbf{v}_i||}{||\mathbf{u}_i||} \tag{6.66}$$

3. If $S_{fi}$ is not less than $\delta$, the next basis vector is computed from

$$\mathbf{e}_{k+1} = \frac{\mathbf{v}_i}{||\mathbf{v}_i||} \tag{6.67}$$

If $S_{fi}$ is less than $\delta$, $\mathbf{u}_i$ is eliminated or constrained. In either case, the above procedure is then repeated for the next $\mathbf{u}_i$, until it has been completed for $\mathbf{u}_m$.

The correction mass calculation is then performed using the remaining columns of the influence coefficient matrix. With the exception of the significance factors and Euclidean norms, all values used in this procedure are complex.

In practical applications, it might be more desirable to combine the non-independent columns of the influence coefficient matrix with the independent columns, rather than simply discarding the former. The result of this operation would be to distribute the correction among all of the originally specified balancing planes while ensuring that the calculation of the correction masses is well behaved. It might also be desirable to evaluate the vector **x** with regard to its effect on this calculation, as mentioned above. This would probably be useful only for cases with marginal significance factors, say between 0.1 and 0.3. This evaluation would involve representing **x** in terms of the basis vectors, $\mathbf{e}_j$, as

$$\mathbf{x} = \sum_{j=1}^{k} \alpha_j \mathbf{e}_j \tag{6.68}$$

where the $\alpha_j$ are the coefficients of the basis vectors. Since the basis vectors are orthonormal, these coefficients can be calculated simply by taking the inner product of **x** with the corresponding basis vector. In general, the more nearly equal in value the $\alpha_j$ are, the less well behaved is the correction mass calculation.

## Constraining the Non-independent Planes

If it is desired to constrain the non-independent balancing planes rather than to eliminate them, a linear programming approach, such as those proposed in [56] and [82] may be used. For these methods, specific constraints must be defined, such as restricting the sizes of the correction masses.

A slightly more arbitrary, though much simpler, means of constraining the non-independent planes would involve combining the influence coefficients for the non-independent planes with those for the independent planes and determining the correction masses in a corresponding manner. As a simple example, consider a two plane case in which the planes are found to be non-independent. The second plane can be constrained very simply by adding the two columns of the influence coefficient matrix together to get a new influence coefficient matrix with a single column, for calculating a correction mass. The resulting correction mass would then be applied, exactly as calculated, in both balancing planes.

If this is done for the case represented by Equations (6.55) and (6.56), the calculated correction mass is given by

$$w = -\frac{(c+1)\left[\left(a^2+b^2\right)d + b\varepsilon_2\right] + db\varepsilon_1 + \varepsilon_1\varepsilon_2}{(c+1)\left[\left(a^2+b^2\right)(c+1) + 2b\varepsilon_1\right] + \varepsilon_1^2} \tag{6.69}$$

where $w$ has a single complex value. Dropping the $\varepsilon^2$ terms and other relatively insignificant terms, specifically $b\varepsilon_2$ and $2b\varepsilon_1$, gives

$$w = -\frac{d}{c+1} - \frac{db\varepsilon_1}{(c+1)\left(a^2+b^2\right)} \tag{6.70}$$

where the second term is clearly much smaller than the first. Thus

$$w \approx -\frac{d}{c+1} \tag{6.71}$$

This correction is then applied to both planes, as indicated in Figure 6.3. If these two planes have approximately the same sensitivity (i.e., $c \approx 1$), then the resulting correction in each plane is about half of the single plane correction illustrated in Figure 6.1.

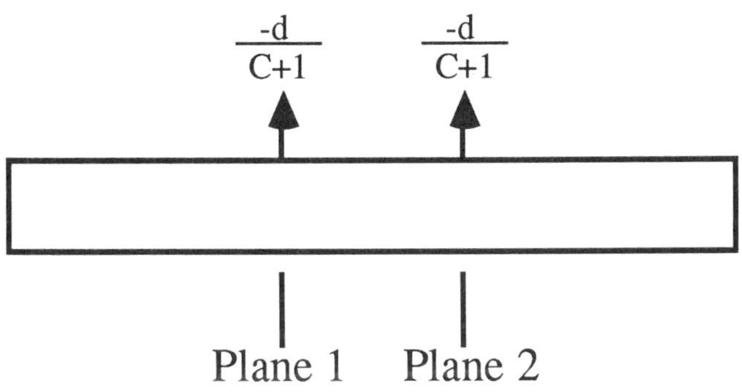


**Figure 6.3** Illustration of Two Plane Correction with the Second Plane Constrained to Have the Same Correction as the First Plane

While this procedure would work well if the two columns of the influence coefficient matrix were generally in phase, it would not be effective if they were generally out of phase. This is apparent from the fact that $w$ becomes infinite as $c$ approaches negative one. Thus, it is necessary to take into account the initial relationship between the columns of the influence coefficient matrix for the independent and non-independent planes. This relationship may be defined, for example, by

$$u_{ij} = \frac{\bar{\mathbf{u}}_i^T \mathbf{u}_j}{\|\mathbf{u}_j\|^2} \qquad \text{for } j = 1,2,\ldots,m' \text{ and } i = m'+1, m'+2, \ldots, m \qquad (6.72)$$

where the subscripts $j$ and $i$ refer to the independent and non-independent balancing planes, respectively, and $m'$ of the original $m$ planes are independent.

The $u_{ij}$ can then be used to modify the columns of the influence coefficient matrix for the independent planes, to include the effect of the non-independent planes, as in

$$\mathbf{u}_j' = \mathbf{u}_j + \sum_{i=m'+1}^{m} u_{ij}\mathbf{u}_i \qquad \text{for } j = 1,2,\ldots,m' \qquad (6.73)$$

where the $\mathbf{u}_j'$ are the modified influence coefficients to be used in the correction mass calculations. Then, given the calculated correction masses for the independent balancing planes, $w_j$ ($j = 1,2,\ldots,m'$), the correction masses for the non-independent planes are calculated from

$$w_i = \sum_{j=1}^{m'} u_{ij} w_j \qquad \text{for } i = m'+1, m'+2, \ldots, m \qquad (6.74)$$

This procedure will tend to result in smaller correction masses at the less sensitive, non-independent planes. Alternatively, this method may be modified to provide larger correction masses at the less sensitive, non-independent planes, or roughly equivalent correction masses at these planes, by replacing Equation (6.72) with Equation (6.75) or (6.76), respectively.

$$u_{ij} = \frac{\bar{\mathbf{u}}_i^T \mathbf{u}_j}{\|\mathbf{u}_i\|^2} \qquad \text{for } j = 1,2,\ldots,m' \text{ and } i = m'+1, m'+2, \ldots, m \qquad (6.75)$$

$$u_{ij} = \frac{\bar{\mathbf{u}}_i^T \mathbf{u}_j}{|\bar{\mathbf{u}}_i^T \mathbf{u}_j|} \qquad \text{for } j = 1,2,\ldots,m' \text{ and } i = m'+1, m'+2, \ldots, m \qquad (6.76)$$

This procedure may be modified in this, or other, ways, according to personal taste and what is found to be most successful for particular applications. However, in order to ensure that the effectiveness of the overall balancing procedure is not being reduced by constraining the non-independent planes in this way, it is necessary that the condition

$$\|\mathbf{u}_j'\| \geq \|\mathbf{u}_j\| \qquad \text{for } j = 1,2,\ldots,m' \qquad (6.77)$$

be satisfied. The effect of constraining, rather than eliminating, the non-independent planes should be to distribute the correction masses over a larger number of balancing planes.

## Numerical Examples

Two test cases were reported by Darlow [27] to demonstrate the balancing plane optimization procedure. The results of these test cases are repeated here, with some errors corrected in the calculations. The data used for these test cases is not actual test data, but rather was generated artificially. For one of these test cases the data was intentionally arranged to provide a somewhat ill conditioned influence coefficient matrix. However, this ill conditioning is not unrepresentative of the level of ill conditioning often encountered in real influence coefficient matrixes.

The data for the test cases are all complex-valued. The influence coefficient matrices for the test cases are rectangular, with four rows and three columns. The corrected results for the test cases are presented in Tables 6.1, 6.2 and 6.3, and illustrated in Figures 6.4, 6.5 and 6.6. The correction masses shown in the tables were calculated using the standard least-squares solution with no weighting factors or iterative solutions.

**Table 6.1** Data and Results for First Test Case (No Plane Elimination

| | Influence Coefficients | | | Uncorrected |
|---|---|---|---|---|
| | Plane 1 | Plane 2 | Plane 3 | Rotor Data |
| Sensor 1 | 1.41 at 45° | 2.24 at 27° | 3.61 at 34° | 3.16 at 72° |
| Sensor 2 | 3.16 at 72° | 4.47 at 27° | 2.24 at 27° | 3.16 at 18° |
| Sensor 3 | 2.83 at 45° | 2.24 at 27° | 5.0 at 37° | 4.12 at 14° |
| Sensor 4 | 3.16 at 18° | 3.61 at 34° | 4.47 at 27° | 5.39 at 68° |

**Correction Masses**

| Plane 1 | Plane 2 | Plane 3 |
|---|---|---|
| 1.39 at −4° | 1.25 at −144° | 0.98 at 168° |

**Significance Factors**

| Plane 1 | Plane 2 | Plane 3 |
|---|---|---|
| 0.331 | 0.502 | 1.0 |

**Table 6.2** Data and Results for Second Test Case Before Plane Optimization

| | Influence Coefficients | | | Uncorrected |
|---|---|---|---|---|
| | Plane 1 | Plane 2 | Plane 3 | Rotor Data |
| Sensor 1 | 1.41 at 45° | 3.61 at 34° | 3.61 at 34° | 3.16 at 72° |
| Sensor 2 | 3.16 at 72° | 2.24 at 27° | 2.24 at 27° | 3.16 at 18° |
| Sensor 3 | 2.83 at 45° | 5.0 at 37° | 5.0 at 37° | 4.12 at 14° |
| Sensor 4 | 3.16 at 18° | 3.61 at 34° | 4.47 at 27° | 5.39 at 68° |

**Correction Masses**

| Plane 1 | Plane 2 | Plane 3 |
|---|---|---|
| 0.87 at 101° | 4.74 at 100° | 5.08 at –87° |

**Significance Factors**

| Plane 1 | Plane 2 | Plane 3 |
|---|---|---|
| 0.466 | 0.110 | 1.0 |

**Table 6.3** Data and Results for Second Test Case After Plane Optimization

| | Influence Coefficients | | Uncorrected |
|---|---|---|---|
| | Plane 1 | Plane 3 | Rotor Data |
| Sensor 1 | 1.41 at 45° | 3.61 at 34° | 3.16 at 72° |
| Sensor 2 | 3.16 at 72° | 2.24 at 27° | 3.16 at 18° |
| Sensor 3 | 2.83 at 45° | 5.0 at 37° | 4.12 at 14° |
| Sensor 4 | 3.16 at 18° | 4.47 at 27° | 5.39 at 68° |

**Correction Masses**

| Plane 1 | Plane 3 |
|---|---|
| 0.51 at 46° | 1.13 at –155° |

**Significance Factors**

| Plane 1 | Plane 3 |
|---|---|
| 0.466 | 1.0 |

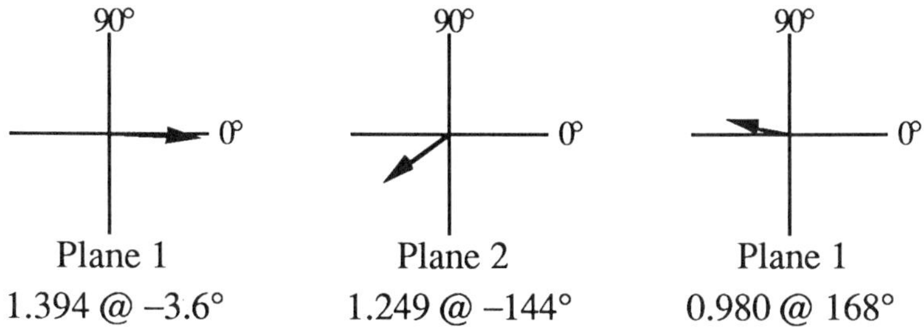


**Figure 6.4** Correction Masses for First Test Case

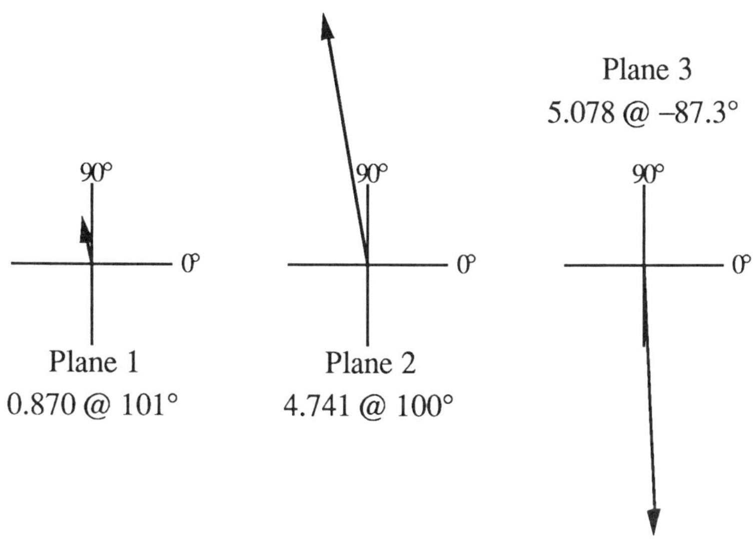


**Figure 6.5** Correction Masses for Second Test Case with Non-Independent Balancing Planes

For the first test case, the influence coefficient matrix was fairly well conditioned, as indicated by the values of the significance factors in Table 6.1. The calculated correction masses, represented in Figure 6.4, shows that while some canceling of the correction masses seems to be in evidence, the amount is fairly small and the total masses are reasonable.

For the second test case, the influence coefficient matrix was not well conditioned, as indicated by the value of the significance factors in Table 6.2 for balancing plane 2. The corresponding correction masses are large and tending to cancel, as illustrated in Figure 6.5. Based on the small significance factor for plane 2, a reduced solution was calculated, eliminating that balancing plane.

Clearly, the reduced solution in Table 6.3 is a substantial improvement over the original solution, as illustrated in Figure 6.6.

**Figure 6.6** Correction Masses for Second Test Case After Optimizing Balancing Plane Selection

There are certainly still some questions to be answered with regard to the practical application of this balancing plane optimization procedure. Specifically, these questions concern the effect on real rotor balancing test data, a refinement of the acceptable level of the significance factor and whether this acceptable level should be related to the order of the influence coefficient matrix.

# Chapter 7
# The Unified Balancing Approach

The Unified Balancing Approach has been designed to incorporate the advantages of both the influence coefficient and modal balancing methods, while eliminating the disadvantages of both methods. That is, the Unified Balancing Approach uses a modal method of applying correction masses in modal sets using data derived in an empirical manner and requiring a minimum of prior knowledge of the dynamics of the rotor. Essentially, the technique involves the calculation of modal trial mass sets. Generally, these modal trial mass sets are determined such that they affect the mode of interest while not having any effect on the lower modes that have already been balanced. However, using the appropriate data, if available from previous tests, a modal trial mass set can be constructed that will have no effect on any general set of modes (hereafter referred to as the unaffected modes), which can include modes above and below the mode being excited. In general, the number of planes required for the modal trial mass set is one more than the number of modes which must not be affected. In addition, the Unified Balancing Approach is not restricted to planar modes. In this chapter, the theoretical development of the Unified Balancing Approach is presented along with the step-by-step procedure required for its implementation.

## Analytical Development

The complex notation representing rotor vibration for the Unified Balancing Approach is the same as that used previously for synchronous vibration and for influence coefficient balancing. To review, the synchronous vibration, as measured in one stationary transverse direction, can be represented as

$$x = |x| \cos(\omega t + \beta) = x_c \cos(\omega t) - x_s \sin(\omega t) \tag{7.1}$$

where

$$x_c = |x| \cos \beta \tag{7.2}$$

$$x_s = |x| \sin \beta \tag{7.3}$$

$|x|$ = vibration amplitude

$\beta$ = leading phase angle

$\omega$ = steady angular speed of shaft rotation

This vibration amplitude may also be represented with the conventional complex notation by

$$\hat{x} = x_c + ix_s \tag{7.4}$$

which is a short form representing

$$x = \text{Real}\left[(x_c + ix_s)\, e^{i\omega t}\right] \tag{7.5}$$

In a similar manner, rotor mass unbalance can be represented in complex form by

$$\hat{u} = u_c + iu_s \tag{7.6}$$

where $u_c$ and $u_s$ are the components of mass unbalance in rotating coordinates. The angle between the stationary coordinate axis along which $x$ is measured and the rotating coordinate axis along which the $u_c$ component is taken is $\omega t$. As virtually all variables in this section are complex, the superior carets are dropped for convenience.

## Calculation of Modal Trial Mass Sets

The first step in the Unified Balancing Approach involves the calculation of a set of complex relationships which define the construction of a modal trial mass set for the mode to be corrected and will not introduce fresh unbalance to the modes which have already been corrected [29, 35]. Hereafter, these complex relationships are referred to as modal trial mass set ratios. A total of $m$ planes in the rotor are used to make up this modal trial mass set so that the number of planes is one greater than the number of modes which are to be unaffected by this modal trial mass set.

A total of $n$ vibration measurements are taken to determine the response of the rotor at the $(m-1)$ critical speeds specified above, with at least one measurement being taken at each of these critical speeds. For a linear rotor system, these vibration measurements, represented by the column vector $\mathbf{x}_w$ of length $n$, in response to masses placed in each of the $m$ individual planes are given by

$$\mathbf{x}_w = \mathbf{A}\mathbf{w} \tag{7.7}$$

where $\mathbf{w}$ is a column vector of length $m$ representing the masses in these planes and $\mathbf{A}$ is an $n$ by $m$ matrix whose elements, $\alpha_{ij}$, are the influence coefficients relating the rotor response for these specified modes to the balancing planes. The $\alpha_{ij}$, are determined empirically through the application of known trial masses in each of the balancing planes, one plane at a time. In the simplest case, a single trial mass is used for each plane and

$$\alpha_{ij} = \frac{x_{ij} - x_{i0}}{T_j} \tag{7.8}$$

where $x_{i0}$ is the $i^{\text{th}}$ vibration reading with no trial masses installed, $x_{ij}$ is the $i^{\text{th}}$ vibration reading with a trial mass installed in the $j^{\text{th}}$ balancing plane, and $T_j$ is a complex value representing the amplitude and angular location of this trial mass Thus, $(m+1)$ test runs, including the run with no trial masses, are required to provide sufficient data to calculate all $\alpha_{ij}$, where $n$ measurements are taken for each test run. The influence coefficients can also be calculated using a measurement error compensation algorithm such as that described in Chapter 6.

In order to fulfill the requirement that the modal trial mass set not affect the $(m-1)$ specified modes, it is necessary that the relation

$$\mathbf{x}_w = \mathbf{A}\mathbf{w} = \mathbf{0} \tag{7.9}$$

be satisfied, where $\mathbf{0}$ represents a column vector of zeroes. The value of $n$ is restricted to

$$n \geq m - 1 \tag{7.10}$$

since at least one vibration reading is required for each of the $(m-1)$ specified modes. To simplify the specification of the modal correction mass sets, a reference plane is chosen from the set of m planes in this modal trial mass set. The choice of the reference plane is arbitrary and is used during the balancing procedure to represent the modal trial mass set. Equation (7.9) can then be rewritten as

$$\mathbf{A'w'} + \mathbf{a}_R w_R = \mathbf{0} \tag{7.11}$$

where $w_R$ represents the mass in the reference plane. In addition $\mathbf{a}_R$ is a column vector formed from that column of the matrix **A** which corresponds to the reference plane. Also, **A'** is a $n$ by $(m-1)$ matrix which is formed by removing the column from **A** that forms $\mathbf{a}_R$, and **w'** is a column vector formed by removing the element $w_R$ from **w**. For example, if the reference plane is plane $k$, then $\mathbf{a}_R$ is a column vector identical with the $k^{\text{th}}$ column of **A**. Also, **A'** is the $n$ by $(m-1)$ matrix which is formed by removing the $k^{\text{th}}$ column from **A**, and **w'** is a column vector, of order $(m-1)$, formed by removing the element $w_R$ from the $k^{\text{th}}$ row of **w**.

For the case when

$$n = m - 1 \tag{7.12}$$

the matrix **A'** is a square, rank $n$, matrix and Equation (7.11) can be solved for **w'** directly to give

$$\mathbf{w'} = -w_R[\mathbf{A'}]^{-1}\mathbf{a}_R \tag{7.13}$$

where, of course, $w_R$ is a scalar quantity. Equation (7.13) can be rewritten as

$$\frac{1}{w_R}\mathbf{w'} = -\,[\mathbf{A'}]^{-1}\mathbf{a}_R = \mathbf{r}_w \tag{7.14}$$

where $\mathbf{r}_w$ is a column vector of order $(m-1)$ containing the complex ratios that relate each of the masses in the modal trial mass set to the mass in the reference plane. If all of the elements of $\mathbf{r}_w$ are real, then the corrections for the particular mode are all located in a single diametral plane along the rotor. However, this will not generally be the case. When this modal trial mass set is applied, the size and angular location of the mass in the reference plane can be chosen arbitrarily, just as an individual trial mass is chosen arbitrarily during influence coefficient balancing. The trial mass in each of the remaining planes is determined by multiplying the reference plane trial mass by the appropriate *modal trial mass ratio*, or element of $\mathbf{r}_w$. This trial mass set will then have no effect on the $(m-1)$ modes specified above. The balancing calculations outlined below are performed using the reference plane trial mass to represent the modal trial mass set and the calculated correction mass is, in actuality, the element of the correction mass set located in the reference plane. The remaining masses in the correction mass set are then calculated from the reference plane correction mass and $\mathbf{r}_w$, as described above.

For the more general case when

$$n > m - 1 \tag{7.15}$$

a weighted least-squares minimization procedure, similar to that described in Chapter 6 for influence coefficient correction mass calculation, is performed. In this case, it is not possible to choose a modal trial mass set such that $\mathbf{x}_w$ in Equation (7.7) is identically equal to zero; that is, with all elements of the vector equal to zero. Instead, the sum of the squares, $s_w$, of the elements of $\mathbf{x}_w$ is minimized where

$$s_w = \sum_{i=1}^{n} | x_{wi} |^2 = \sum_{i=1}^{n} \bar{x}_{wi} x_{wi} = \bar{\mathbf{x}}_w^T \mathbf{x}_w \tag{7.16}$$

The $x_{wi}$ are the elements of $\mathbf{x}_w$ and the superscript bar denotes the complex conjugate. Following the procedure of Lund and Tonneson [87], $s_w$, which is real-valued, is minimized by

$$\frac{\partial s_w}{\partial \bar{w}_j} = 0 = \sum_{i=1}^{n} x_{wi} \frac{\partial \bar{x}_{wi}}{\partial \bar{w}_j} = \sum_{i=1}^{n} x_{wi} \bar{\alpha}_{ij} \qquad \text{for } j = 1,2,\ldots,m \tag{7.17}$$

where the reference plane is again excluded. Rewriting Equation (7.7) as

$$\mathbf{x}_w = \mathbf{A}'\mathbf{w}' + w_R \mathbf{a}_R \tag{7.18}$$

and substituting the result into Equation (7.17) gives, in matrix form

$$\bar{\mathbf{A}}'^T \left( \mathbf{A}'\mathbf{w}' + w_R \mathbf{a}_R \right) = 0 \tag{7.19}$$

where $\bar{\mathbf{A}}'^T$ is the complex conjugate transpose of the matrix $\mathbf{A}'$. Solving Equation (7.19) for $\mathbf{w}'$ gives

$$\mathbf{w}' = - w_R \left( \bar{\mathbf{A}}'^T \mathbf{A}' \right)^{-1} \bar{\mathbf{A}}'^T \mathbf{a}_R \tag{7.20}$$

Dividing by $w_R$, which is a scalar, gives the solution for the ratio vector $\mathbf{r}_w$ as

$$\mathbf{r}_w = - \left( \bar{\mathbf{A}}'^T \mathbf{A}' \right)^{-1} \bar{\mathbf{A}}'^T \mathbf{a}_R \tag{7.21}$$

This equation represents the choice of the modal trial mass set which has the minimum effect on the selected modes, in a least squares sense.

The anticipated response to this modal trial mass set for each of the vibration readings is calculated as

$$\mathbf{x}'_w = \frac{1}{w_R}\mathbf{x}_w = \mathbf{A}'\mathbf{r}_w + \mathbf{a}_R \tag{7.22}$$

where $\mathbf{x}'_w$ is the vector of resultant vibration readings for a modal trial mass set with a unit trial mass, at zero degrees, in the reference plane.

Although the modal trial mass set calculated from Equation (7.21) has the least overall effect on the $n$ vibration readings, the effect on the individual vibration readings may be widely variable. If it is preferred to minimize the maximum effect on a vibration reading (i.e., the largest element of $\mathbf{x}_w$ or $\mathbf{x}'_w$), an iteration procedure may be used, which is again similar to that described in Chapter 6 for influence coefficient correction mass calculation. In general, this procedure tends to equalize the magnitude of the elements of $\mathbf{x}'_w$ at the expense of the root mean square average of these elements.

To perform this procedure, Equation (7.16) is rewritten as

$$s_{wk} = \sum_{i=1}^{n} \varepsilon_{ik} \left| x'_{wi} \right|_k^2 = \text{minimum} \qquad \text{for } k = 0,1,2,... \tag{7.23}$$

where $k$ is the iteration number and $\varepsilon_{ik}$ is a weighting factor which is given by

$$\varepsilon_{ik} = \prod_{p=0}^{k-1} \frac{\left| x'_{wi} \right|_p}{(\text{rms})_p} = \frac{\left| x'_{wi} \right|_{k-1}}{(\text{rms})_{k-1}}\, \varepsilon_{i,k-1} \tag{7.24}$$

where $\prod$ denotes a multiplication series, $(\text{rms})_p$ is the real-valued, root mean square average of the elements of the vector $\mathbf{x}'_w$ from the $p^{\text{th}}$ iteration of the modal trial mass set calculation, and $\varepsilon_{ik}$ is real. Thus

$$(\text{rms})_p = \sqrt{\frac{\sum_{i=1}^{n} \left| x'_{wi} \right|_p^2}{n}} \tag{7.25}$$

The first iteration ($k = 0$) is identical to the least-squares calculation described above in Equation (7.21), which is obtained for

$$\varepsilon_{i0} = 1 \qquad \text{for } i = 1,2,\ldots,n \tag{7.26}$$

A result analogous to Equation (7.21) including the effect of the weighting factor is found in a similar manner as above. The final result is

$$\mathbf{r}_{wk} = -\left(\bar{\mathbf{A}}'^T \mathbf{E}_k \mathbf{A}'\right)^{-1} \bar{\mathbf{A}}'^T \mathbf{E}_k \mathbf{a}_R \tag{7.27}$$

where $\mathbf{r}_{wk}$ is the $k$th iteration of $\mathbf{r}_w$, and $\mathbf{E}_k$ is an $n$ by $n$ diagonal matrix in the form

$$\mathbf{E}_k = \begin{bmatrix} \varepsilon_{1k} & 0 & 0 & 0 \\ 0 & \varepsilon_{2k} & 0 & 0 \\ 0 & 0 & \cdots & \cdots \\ 0 & 0 & \cdots & \varepsilon_{nk} \end{bmatrix} \tag{7.28}$$

A convergence limit can be specified for the iteration procedure such that when

$$\frac{\left|\, |x'_{wi}|_k - |x'_{wi}|_{k\text{-}1} \right|_{\max}}{(\text{rms})_k} \leq \text{convergence limit} \tag{7.29}$$

the iterations are considered to have converged.

It should be noted that the results of this least-squares procedure, with or without the weighting factors, are dependent on the scaling of the vibration readings. Thus, to obtain meaningful results, it is generally advisable that the vibrations be measured in the same units and have the same order of magnitude. By the same token, if it is desirable for the effect of the modal trial mass set on a specific vibration reading, or readings, be less than the effect on the remaining vibration readings, a magnification factor can be applied to the corresponding elements of the weighting matrix before the application of Equation (7.27). In this way, a considerable amount of flexibility can be introduced into the procedure for calculating modal trial mass sets. Such a magnification factor is applied for a particular vibration reading by multiplying the appropriate row of the matrix $\mathbf{E}_k$ by this factor.

## Calculation of Modal Correction Mass Sets

After the modal trial mass set has been calculated for a particular mode to be balanced, vibration readings are taken at that mode for the uncorrected rotor system, and then with a modal trial mass set installed. A column vector, of length $N$, of modal influence coefficients is calculated, where $N$ is the number of vibration readings taken at the mode of interest. This is done in a manner exactly analogous to that used for calculating influence coefficients as described above for standard influence coefficient balancing. The only difference is that, in this case, the trial mass located in the reference plane is used to represent the entire modal trial mass set even though it is not used independently.

Similarly, the balancing calculations are exactly analogous to single-plane influence coefficient balancing, where only the reference plane correction mass is calculated directly and the remaining correction masses are determined using the appropriate modal mass set ratios. Since this balancing calculation always involves only a single balancing plane, the influence coefficient balancing solution equations can be simplified considerably.

The residual vibration of the rotor system, $\mathbf{x}$, after the addition of a modal correction mass set, is given as

$$\mathbf{x} = \mathbf{x}_0 + w_m \mathbf{a}_m \tag{7.30}$$

where $\mathbf{x}$ is a column vector of length $N$, with elements $x_{ij}$; $\mathbf{x}_0$ is a similar column vector representing the vibration readings for the uncorrected rotor at the mode of interest; $\mathbf{a}_m$ is the modal influence coefficient vector; and $w_m$, which is scalar, is the element of the modal correction mass set located in the reference plane.

The aim of the rotor balancing procedure is to specify $w_m$, and thus to specify all elements of the modal correction mass set, so as to minimize $\mathbf{x}$. For $N$ equal to one, Equation (7.30) can be solved directly to give

$$w_m = - x_0 / \alpha_m \tag{7.31}$$

where $x_0$ and $\alpha_m$ are now scalars.

For the more general case, when $N$ is not equal to one, a least-squares minimization procedure is performed similar to that described above for the calculation of the modal trial mass sets. This procedure is somewhat simplified here, since only vectors and scalars are involved.

In this case, the sum of the squares of the residual amplitudes, $s$, which is again real-valued, is minimized where

$$s = \sum_{i=1}^{N} |x_i|^2 = \sum_{i=1}^{N} x_i \bar{x}_i \tag{7.32}$$

The partial derivative of $s$ with respect to $\overline{w}_m$ is now set equal to zero to give

$$\frac{\partial s}{\partial \overline{w}_m} = \sum_{i=1}^{N} x_i \frac{\partial \overline{x}_i}{\partial \overline{w}_m} = \sum_{i=1}^{N} x_i \, \overline{\alpha}_{mi} = 0 \tag{7.33}$$

where $\overline{\alpha}_{mi}$ is the complex conjugate of the $i$th element of the vector $\mathbf{a}_m$. Substituting Equation (7.30) into Equation (7.33) results in

$$\sum_{i=1}^{N} \overline{\alpha}_{mi} \left(x_{i0} + \alpha_{mi} w_m\right) = \overline{\mathbf{a}}_m^T \mathbf{x}_0 + \overline{\mathbf{a}}_m^T \mathbf{a}_m w_m = 0 \tag{7.34}$$

where $x_{i0}$ is the $i$th element of the vector $\mathbf{x}_0$. Solving Equation (7.34) for $w_m$ gives

$$w_m = -\frac{\left(\overline{\mathbf{a}}_m^T \mathbf{x}_0\right)}{\left(\overline{\mathbf{a}}_m^T \mathbf{a}_m\right)} \tag{7.35}$$

where $\overline{\mathbf{a}}_m^T$ is the complex conjugate transpose of the vector $\mathbf{a}_m$ and the result is a division of scalar quantities.

In a similar fashion to that described above for the calculation of the modal trial mass sets, an iterative weighted least-squares minimization procedure can be used for calculating $w_m$. In this case Equation (7.32) is rewritten as

$$s_k = \sum_{i=1}^{N} \varepsilon_{ik} \left| x_i \right|_k^2 = \text{minimum} \qquad \text{for } k = 0,1,2,\ldots \tag{7.36}$$

where $k$ is again the iteration number and $\varepsilon_{ik}$ is a real-valued weighting factor given by

$$\varepsilon_{ik} = \prod_{p=0}^{k-1} \frac{\left| x_i \right|_p}{(\text{rms})_p} = \frac{\left| x_i \right|_{k-1}}{(\text{rms})_{k-1}} \varepsilon_{i,k-1} \tag{7.37}$$

where $(\text{rms})_p$ is real and is given by

$$(\text{rms})_p = \sqrt{\frac{\sum_{i=1}^{N} |x_i|_p^2}{N}} \tag{7.38}$$

The first iteration ($k = 0$) is identical to the least-squares calculation described above, resulting in Equation (7.35), which is obtained for

$$\varepsilon_{i0} = 1 \qquad \text{for } i = 1,2,\ldots,N \tag{7.39}$$

Minimizing the value of Equation (7.36) in the same manner as was done for Equation (7.32) yields

$$w_{mk} = -\frac{\left(\bar{\mathbf{a}}_m^T \mathbf{E}_k \mathbf{x}_0\right)}{\left(\bar{\mathbf{a}}_m^T \mathbf{E}_k \mathbf{a}_m\right)} \tag{7.40}$$

where $w_{mk}$ is the $k^{\text{th}}$ iteration for $w_m$ and $\mathbf{E}_k$ is defined in Equation (7.28). The discussion involving the scaling of vibration readings presented in the previous section applies to the use of Equations (7.35) and (7.40), as well.

## Calculation of Influence Coefficients

The procedure for calculating and optimizing influence coefficients is essentially the same whether individual trial mass or modal trial mass sets are used. The only difference is that the complex value of the reference plane trial mass only is used to represent the entire modal trial mass set, while the individual trial mass values are each used when calculating individual influence coefficients.

In general, placing a known trial mass, $T_j$, at a known angle in plane $j$ results in a new $i^{\text{th}}$ vibration reading, $x_{ij}$, giving the corresponding influence coefficient as

$$\alpha_{ij} = \frac{x_{ij} - x_{i0}}{T_j} \tag{7.41}$$

where $x_{i0}$ is the $i^{\text{th}}$ vibration reading for the undisturbed rotor; and $j = 1,\ldots, m$ for individual trial masses or $j = 1$ for a modal trial mass set. However, there are certain errors inherent in the measurement of real vibration data. Lund and Tonneson [87] defined a procedure by which a second trial mass is used in the same plane and the two sets of trial mass data are used to optimize the influence coefficients and reduce the effect of these inherent measurement errors. The equations derived by Lund to perform this optimization procedure are reproduced here and the derivations are presented in Chapter 6. The optimized influence coefficients are calculated as

$$\alpha_{ij} = \frac{1}{|T_j|^2 + |T_j'|^2}\left[\overline{T}_j\left(x_{ij} - x_{i0} - \Delta x_{i0}\right) + \overline{T}_j'\left(x_{ij}' - x_{i0} - \Delta x_{i0}\right)\right] \tag{7.42}$$

where the primed quantities refer to the second trial mass run and $\Delta x_{i0}$, the measurement error for the undisturbed rotor vibration data, is given by

$$\Delta x_{i0} = \frac{\displaystyle\sum_{j=1}^{m} \frac{\overline{T}_j - \overline{T}_j'}{|T_j|^2 + |T_j'|^2}\left[T_j\left(x_{ij}' - x_{i0}\right) - T_j'\left(x_{ij} - x_{i0}\right)\right]}{1 + \displaystyle\sum_{j=1}^{m} \frac{|T_j - T_j'|^2}{|T_j|^2 + |T_j'|^2}} \tag{7.43}$$

When calculating modal correction masses, where $m = 1$, Equation (7.43) reduces to

$$\Delta x_{i0} = \frac{\left(\overline{T}_1 - \overline{T}_1'\right)\left[T_1\left(x_{i1}' - x_{i0}\right) - T_1'\left(x_{i1} - x_{i0}\right)\right]}{|T_1|^2 + |T_1'|^2 + |T_1 - T_1'|^2} \tag{7.44}$$

The values for the $x_{i0}$ are subsequently modified to remove the measurement error by adding the $\Delta x_{i0}$ before calculating the modal correction mass set from Equation (7.31), Equation (7.35) or Equation (7.40). In addition, it should be noted that for the special case when

$$T_j' = -T_j \tag{7.45}$$

Equation (7.40) reduces to

$$\alpha_{ij} = \frac{x_{ij} - x_{ij}'}{2T_j} \tag{7.46}$$

which is independent of $\Delta x_{i0}$. However, even for this special case, $x_{i0}$ is still modified by adding the correction $\Delta x_{i0}$.

## Procedure for Application of the Unified Balancing Approach

The procedure for applying the Unified Balancing Approach, illustrated in the flowchart in Figure 7.1, is described here to give a clearer understanding of the process [29]. It should be kept in mind that the Unified Balancing Approach is inherently applications oriented. That is, this approach is designed to handle a number of non-ideal, but not uncommon, conditions including the occurrence of non-planar or distorted mode shapes and the existence of measurement error and finite vibration data resolution. A computer program is generally required to perform the necessary calculations.

The basic procedure for implementing the Unified Balancing approach is as follows (the paragraph numbers correspond to the step numbers on the flowchart in Figure 7.1):

1. The balancing specifications for the rotor system are defined. These specifications include the number, types (e.g. displacement, velocity, acceleration or force) and calibration factors of the vibration sensors to be used and the specific sensors which are most sensitive for each of the modes to be balanced; the number of balancing planes and balancing mass holes specifications for each plane, as well as the specific planes which comprise each of the modal trial mass sets; the angular location of the trial masses and whether additional trial mass runs for reduction of measurements errors are to be used; the number of modes to be balanced and the speed at which vibration data are to be taken for each mode (these values may be changed by the operator during the balancing procedure, if necessary); previously measured modal influence coefficients for this specific rotor or class of rotors, if available; the number of weighted least-squares minimization iterations to perform if $n > (m - 1)$ or $N > 1$ (the convergence limit is generally taken to be 0.001); adjacent sensor/plane combinations for use in reciprocity calculations, if appropriate, as discussed in Chapter 9; sensor pairs to be used for extraction of forward precession whirl, if appropriate; and any balancing specifications to be applied in steps seven and eight. This is the only step in this balancing procedure which requires engineering-level decisions to be made. The remainder of the steps may be handled by a moderately skilled operator. Thus, this balancing procedure is well suited for application to production balancing, where step one must be done only once.

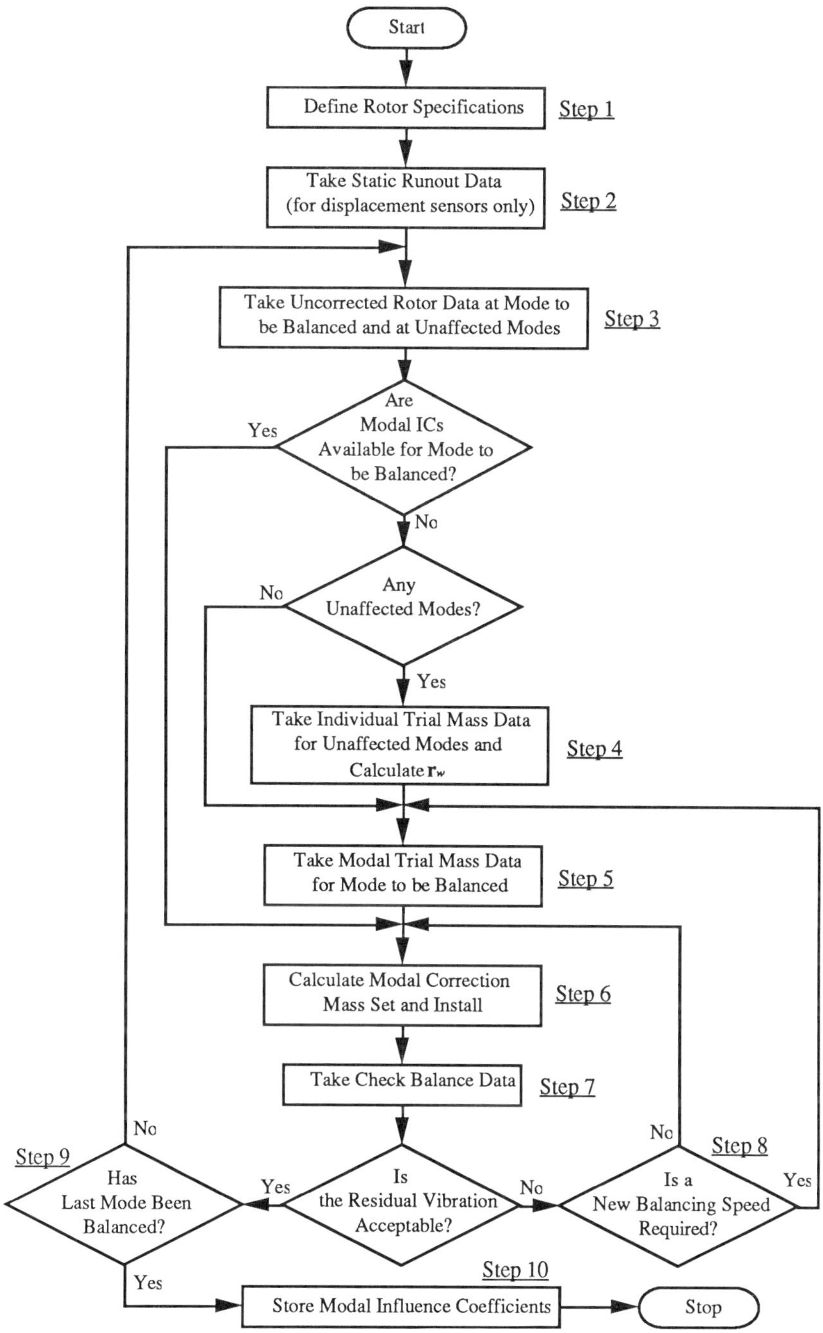


**Figure 7.1** Flowchart of Unified Balancing Approach Procedure

2. The rotor is run at a very slow speed, to ensure that no dynamic response is present, and readings are taken from displacement measuring sensors only. These readings are referred to as static runout data and are due to rotor surface eccentricity at the sensor target and other sources of synchronous data, such as an initial bend in the rotor, which are not caused by rotor mass unbalance. In general, this static runout data is independent of the rotational speed of the rotor. The true vibratory response of the rotor at other rotational speeds is found by subtracting the static runout from the measured vibration readings. If none of the sensors measure rotor displacement, this step may be bypassed.

3. The undisturbed rotor (i.e., no trial masses installed) is run, and vibration readings are taken for the mode to be balanced and for all modes which are not to be affected by the balancing procedure for this mode, at corresponding specified speeds. If previously measured modal influence coefficients are available for the mode to be balanced, the balancing procedure is continued with step six. If this is not the case, but there are no unaffected modes, as is generally true when balancing the first rotor mode, the balancing procedure is continued with step five. Otherwise, step four follows.

4. Individual trial masses are installed, one at a time, in each of the planes to be used for balancing the mode of interest and vibration readings are taken at the specified data speed for each of the unaffected modes. If reciprocity is used, only one balancing plane is used here, as explained in Chapter 9. Either one or two trial masses are used for each balancing plane, as specified in step one. The complex modal trial mass ratios (i.e., the elements of the vector $\mathbf{r}_w$) are then calculated.

5. A modal trial mass set is installed, which consists of a single mass if there are no unaffected modes, and vibration readings are taken at the specified data speed for the mode to be balanced. A second modal trial mass set is installed and another set of vibration readings are taken, if so specified in step one. The modal influence coefficients for the mode being balanced are then calculated, based on this modal trial mass data and the uncorrected rotor data from step three.

6. The modal correction mass set is calculated for the mode being balanced, based on the modal influence coefficients and the uncorrected rotor data from step three, modified to compensate for measurement error if two modal trial mass sets have been used for this mode. The operator then installs this modal correction mass set.

7. Another set of vibration readings is taken at the specified data speed for the mode being balanced to determine whether sufficient improvement of

the rotor response has been achieved by this balancing procedure. The measured residual vibration is either displayed to the operator, for the operator to evaluate, or is compared to previously specified balancing criteria. If the measured residual vibration is determined to be acceptable, the balancing procedure is continued with step nine; otherwise step eight follows.

8. If additional improvement in the measured residual vibration is required, and if it is determined that this improvement can be achieved using the same balancing speed and the same modal influence coefficients, the balancing procedure is continued with step six. If, however, it is determined that a more sensitive balancing speed (i.e., nearer to the critical speed) is required, or that updated modal influence coefficients are needed (e.g., due to non-linearities in the rotor system), the balancing procedure is continued with step five. A balancing speed nearer to the relevant critical speed may also be required to isolate the mode being balancing from other modes.

9. If one or more modes remain to be balanced, the mode just balanced is added to the list of unaffected modes, the next mode to be balanced is chosen as specified in step one, and the balancing procedure continues with step three.

10. After the last mode has been balanced, the modal influence coefficients which have been measured may be stored for later use in re-balancing the same rotor or in balancing other rotors of the same type. This completes the balancing procedure for this rotor.

It should be noted that, when using measured modal influence coefficients to apply to a class of rotors, it may be necessary to measure the modal influence coefficients for several of these rotors and statistically average these values, as in reference [77], to get a set of modal influence coefficients which can be applied with good results to a whole class of rotors.

The Unified Balancing Approach has some apparent advantages over both modal and influence coefficient balancing.

1. No prior knowledge of the vibration modes of the rotor to be balanced is required. The necessary information is determined empirically in the course of balancing.

2. As with modal balancing, the modes can be balanced individually, while not affecting the balance of other modes that have already been balanced.

3. In particular cases where the lower modes are more lightly damped than the higher modes, the use of modal trial mass sets allows the application

of substantial trial masses at the insensitive modes, while not aggravating the response of the lower modes. It should be noted that individual trial mass runs that are required for the Unified Balancing Approach always involve modes that have already been balanced. Therefore, the vibration readings for these individual trial mass runs can always be taken at speeds close to the critical speeds so that small trial masses can be used with substantial effect, while the rotor is not prevented from negotiating lower critical speeds.

4. Using the Unified Balancing Approach, a smaller number of runs is required at the highest speeds of the rotor than would be required for influence coefficient balancing with individual trial mass runs. For rotors which obey the principle of reciprocity, and for which closely adjacent planes and sensors are available, the total number of individual trial mass runs can also be substantially reduced, as discussed in Chapter 9.

5. In many cases, the balancing of a particular mode begins well below the corresponding critical speed, and the response at the balancing speed is reduced to the point where no additional improvement can be made due to limited vibration pickup sensitivity, but the critical speed can still not be negotiated. Only a single trial mass run is required at a higher balancing speed using the Unified Balancing Approach. In contrast, using influence coefficient balancing, a complete new set of trial mass runs is required at the higher speed. Thus, use of the Unified Balancing Approach can result in a substantial reduction in the total number of balancing runs.

6. Production balancing of a series of identical rotors can be greatly simplified through the use of empirically determined and statistically averaged standard modal influence coefficients. In this way, the trial mass runs become few, or even nonexistent, for a production balancing situation.

The principal disadvantages of the Unified Balancing Approach lie in the complexity of the calculations, which are more involved than those of modal balancing; and the somewhat maze-like procedure, which is certainly not as straightforward as that of influence coefficient balancing. These potential disadvantages necessitate the availability and use of a carefully constructed computer program which will guide the engineer and technician through the balancing procedure, with sufficient intelligence to handle any reasonable situation, while performing all of the necessary calculations.

# Chapter 8
# Experimental Comparisons of the Various Methods

In order to evaluate the true merits of any balancing method, it is necessary to apply it to actual hardware. To be credible, a balancing method must have, in addition to theoretical appeal, demonstrated practical appeal. That is, it is necessary to demonstrate both effectiveness and usability in an actual hardware application. Consequently, modal balancing, influence coefficient balancing and the Unified Balancing Approach were compared experimentally in a test program involving a supercritical power transmission shaft test rig.

## Test Rotor System

The test rotor system provided a drive system capable of speeds to 3,600 rpm, a pair of gearboxes, each with approximately 5.7:1 speed ratio, and a high-speed test shaft, which could be driven at speeds up to 20,000 rpm and would pass through several bending critical speeds in this speed range. The two gearboxes were arranged so that their high-speed and low-speed output shafts could be connected in a "four-square" closed loop arrangement into which a torque could be locked. In this way, power levels up to 2,500 hp could be simulated. The test rig, setup for testing a torsionally unloaded shaft, is shown in Figure 8.1 without the damper. For the purposes of the tests described herein, the test shaft was run in the unloaded configuration [34].

The test shaft was a hollow shaft 3.66 m (12 ft) long and 7.62 cm (3 in.) in diameter with a 3.175 mm (0.125 in.) wall thickness. The shaft had seven balancing rings and was supported at both ends by disk-type flexible couplings. One of the couplings was attached to a hard-mounted spindle, while the other was attached to a stub-shaft that was supported by a squeeze-film damper.

An undamped critical speed analysis of the test rig, including the damper shaft, was conducted. Figure 8.2 shows the predicted mode shapes and critical speeds for the first five modes. It should be noted that for all of these modes there is some significant amplitude at the location of the damper, as represented by bearing number 2. This is important to achieve the necessary dissipation of vibration energy at the damper.

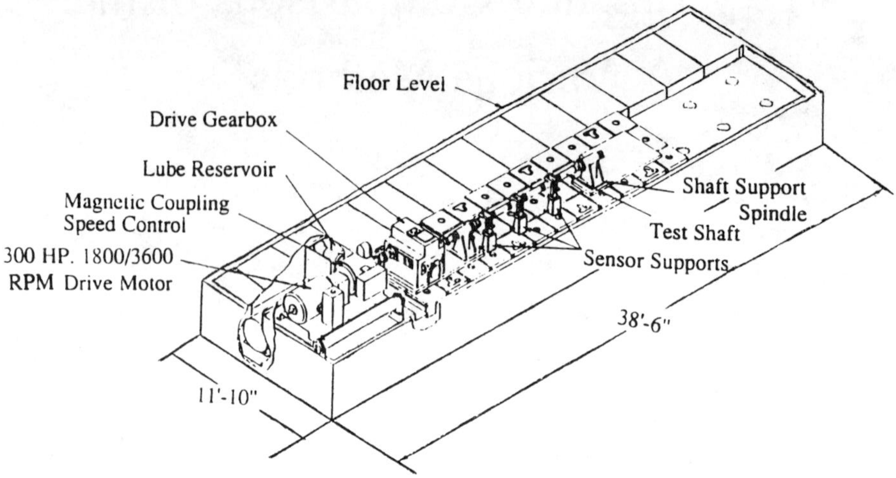


**Figure 8.1** Sketch of Supercritical Shaft Test Facility Prior to Damper Installation

Next, a damped natural frequency and unbalanced response analysis was conducted for the test rig in order to optimize the damping at each critical speed. The objective of the study was to find that combination of practical support properties, stiffness and damping, which maximized the rotordynamic stability of the test rig. The predicted damped critical speeds for the first, third, fourth and fifth modes were 853, 3383, 7500, and 13,200 rpm, respectively. The predicted damped mode shapes were essentially identical to the undamped mode shapes presented in Figure 8.2. The damped natural frequency analysis indicated that the second mode would be critically damped. The measured first, third and fourth critical speeds were approximately 940, 3,500, and 7,600 rpm, respectively.

Test shaft vibration was measured by means of eddy-current-type displacement probes located adjacent to the balancing rings and at the ends of the damper. A sketch indicating the location and reference number of each of these displacement probes is presented in Figure 8.3. The probe or sensor numbers referred to in this chapter correspond to the reference numbers in Figure 8.3. A photograph of the completely assembled test rig, including the damper and displacement probes, is presented in Figure 8.4. The diagram in

Figure 8.5 illustrates the relative locations of the displacement probes and the phase reference sensor and the orientation of the correction mass holes.

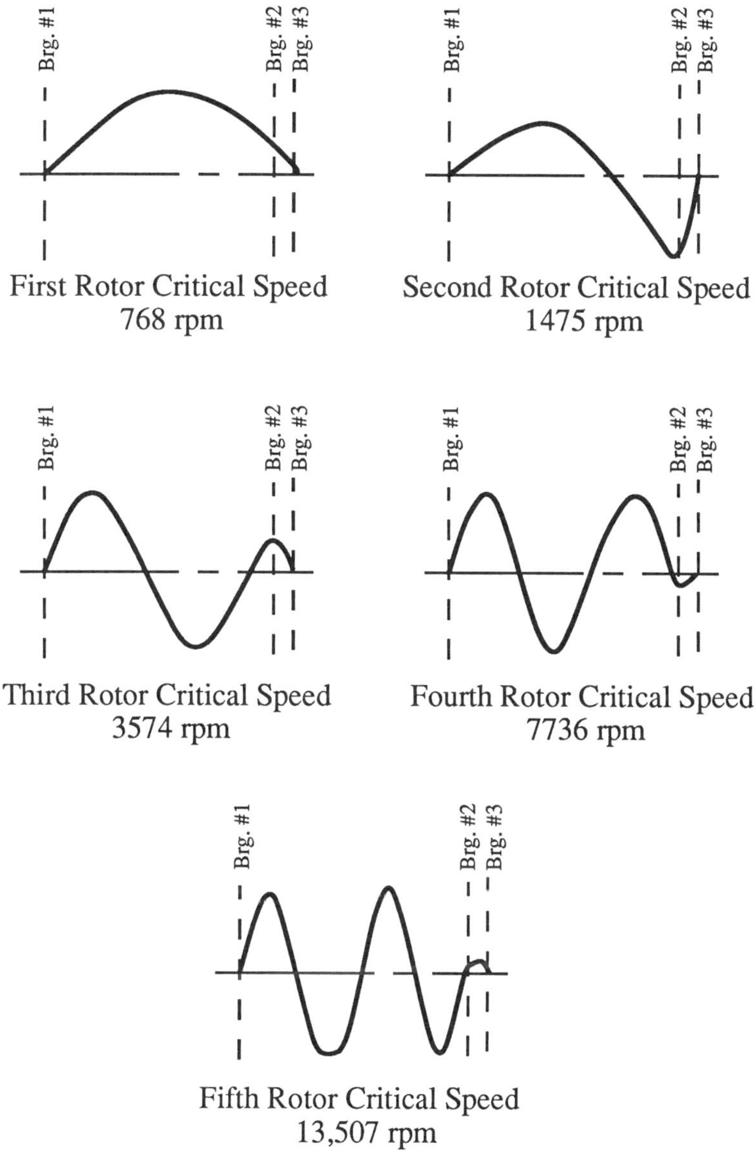


**Figure 8.2** The Mode Shapes for the First Five Critical Speeds of the Supercritical Shaft Test Rig

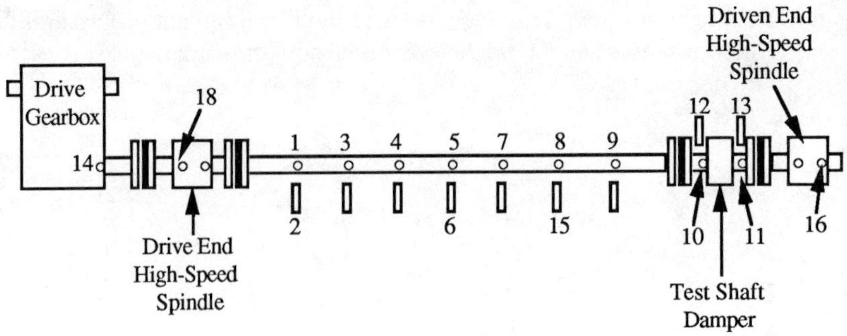


**Figure 8.3** Sensor Layout for Supercritical Shaft Test Rig

**Figure 8.4** Photograph of Supercritical Shaft Test Rig

During the tests, measurements of synchronous test shaft vibration were extracted from the displacement probe signals through the use of a minicomputer controlled data acquisition system. The data acquisition system included a twenty-channel scanner, tracking filter, phase meter, digital voltmeter and tachometer. The minicomputer served the dual purpose of controlling the data acquisition system and analyzing the acquired data. Sophisticated influence coefficient balancing software installed on the minicomputer was used during the test.

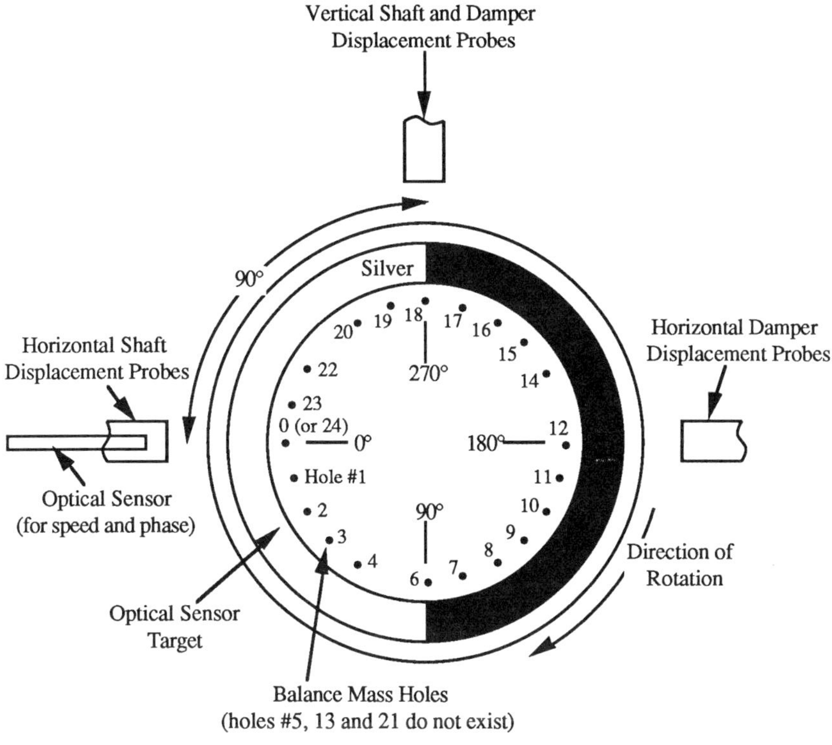


**Figure 8.5** Sketch of Radial Arrangement of Sensors and Correction Mass Hole Locations for Supercritical Shaft Test Rig

Two damper types were used at different stages of the test program. A squeeze-film damper was used during the influence coefficient and modal balancing tests; and an elastomer damper was used during the Unified Balancing Approach tests. A split sketch showing the two damper designs is presented in Figure 8.6. The response of the system with the elastomer damper did not differ from that with the squeeze-film damper except for a slight change in the critical speeds. The shapes of the resonant peaks and the sensitivity of the test rig to unbalance was not noticeably affected by the change in damper configuration. Thus, the type of damper is not a significant factor in the comparisons of balancing test results. Additional details of the design and analysis of the test rig, as well as a more complete description of the hardware, can be found in references [32] and [34]. The test procedures and results are presented below in considerable detail in order to provide a clear picture of the process of flexible rotor balancing.

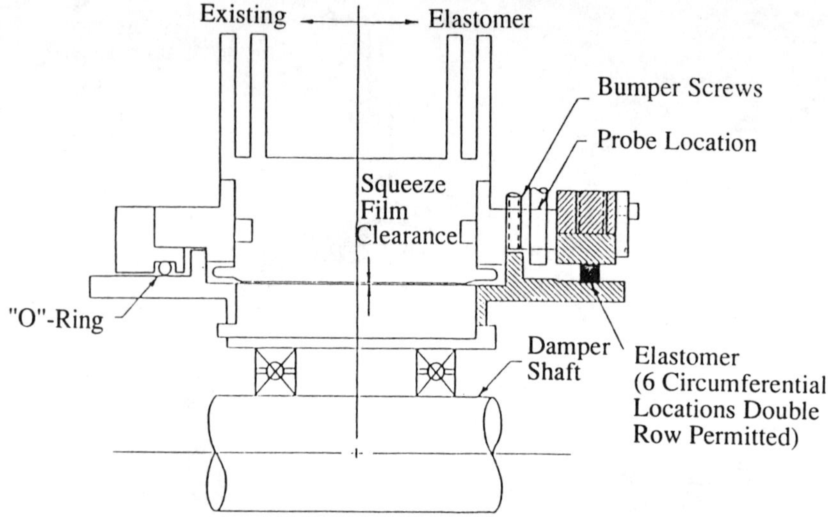


**Figure 8.6** Schematic Diagram of Supercritical Shaft Test Rig Damper Showing Both Squeeze-Film and Elastomer Configurations

## Influence Coefficient Balancing Tests

The first set of balancing tests involved the use of an influence coefficient balancing procedure. An on-line computer program was used to handle data acquisition and balancing calculations. The balancing calculation portion of the program was based on the work of Lund and Tonneson[6].

Observation of the shaft at low speeds in the vicinity of 350 rpm indicated that the shaft had some possible static runout and initial bend. This static runout data was measured and subsequently subtracted, in a vector manner, from the whirl data measured at the balancing speeds to obtain net vibration data. All the subsequent whirl data presented in this chapter, therefore, relate to the net vibration, in amplitude and phase, which has been calculated in this manner.

Some preliminary balancing tests had already been performed with the trial shaft when some instrumentation problems were encountered. This discussion concerns the balancing operation subsequent to restarting the trial, although certain previously determined influence coefficients were applied. For this reason the application of the influence coefficient technique near the first and third critical speeds was probably shorter than would normally be the case, since no trial runs were needed to calculate the influence coefficients. On the other hand, subsequent trial results suggest that the system may have become less sensitive to unbalance, so that the previously derived influence coefficients may not have been entirely reliable in the new tests.

## Balancing at the First Critical Speed

The first critical speed was later established experimentally to be approximately 950 rpm, but the first balancing attempts were made at 901 rpm, at which speed the whirl could be expected to have a large component from the first mode of the shaft. The net initial whirl of the shaft at 901 rpm, as recorded by sensor 5, was measured. Using the influence coefficient, relating plane 4 and sensor 5, which was determined during the preliminary trials, a correction mass was calculated. The predicted correction was determined both in magnitude (grams) and angular location (degrees) around the rotor from position 0 in Figure 8.5. Second, as there is no hole for the insertion of a balance mass at the calculated angle, the mass was distributed between the adjacent holes by the influence coefficient balancing program.

The result of these corrections suggested that the balance mass had markedly under-corrected the unbalance. The amplitude of whirl at sensor 5 was only reduced by about 40 percent. As was remarked earlier, the shaft seemed to be less sensitive to unbalance at 901 rpm than it had been during the preliminary trials, when data was collected for the calculation of the influence coefficient. This change in dynamic response may well have been caused by a change in the operating characteristics of the damper.

The residual whirl data were used to predict the trim balance correction. Due to the apparent decrease in shaft sensitivity, indicated above, the actual correction used was larger than the one suggested by the influence coefficient calculation. The trim correction produced a further 75 percent decrease in whirl amplitude in plane 5, representing an overall reduction of approximately 85 percent from the original whirl amplitude.

Although at this stage it was possible to run the shaft through its first critical speed, the amplitude at this speed was still undesirably high. Consequently, a further trim correction was calculated through the influence coefficient program using the whirl data after the most recent trim correction. Once more, because of the risk of underestimating the size of the correction mass, the actual masses used were greater than those suggested. The residual whirl amplitude showed a further decrease by 60 percent. Indeed, the overall reduction in whirl amplitude achieved was about 93 percent. At this stage the shaft was very well behaved when running through its first critical speed.

## Balancing at the Third Critical Speed

The test shaft was now run up to about 3400 rpm, but it was not possible to run through the third critical speed at about 3460 rpm, due to excessive whirl. Recall that the second mode of the test system was critically damped, so that there was no resonant whirl at the second critical speed. Consequently, a two-speed balance was conducted at 917 rpm and 3200 rpm using whirl data measured by sensors 3, 5, and 8. The balancing planes selected were at locations 2, 4, and 6 in Figure 8.3. The initial whirl of the shaft, after balancing near the first critical speed was measured at the new balancing speeds. These data were used, together with the values for the appropriate influence coefficients (relating planes 2, 4, and 6 to sensors 3, 5, and 8 at the speeds of 917 rpm and 3200 rpm), to calculate balance corrections. The residual whirl produced by this set of corrections showed significant reductions. Indeed, the resulting whirl amplitudes were reduced about 50 percent and 70 percent, respectively, at the first and third critical speeds. The amplitudes of whirl at the first and third critical speeds were low enough at this point that no trim correction was necessary and none was performed.

It should be noted that, out of academic interest, a trim correction was, in fact, calculated on the basis of this residual whirl and the suggested corrections were as follows:

| | |
|---|---|
| Plane 2 | 1.426 grams at −78.7° |
| Plane 4 | 2.105 grams at 100.6° |
| Plane 6 | 1.586 grams at −82.3° |

This is a result which will be discussed further below, in the context of modal balancing. Effectively, however, this distribution is orthogonal to the first and third modes of the shaft and suggests that the residual whirl is mainly in the higher modes of the shaft, especially in the fourth mode. There is not much point in attempting to correct this residual whirl by means of measurements at 917 and 3200 rpm. Much greater accuracy and sensitivity could be expected with measurements made at speeds closer to the fourth critical speed. It is also interesting to note that this distribution is approximately located in one diametral plane through the rotor oriented at 100 degrees from the 0 hole location. By contrast, the initial corrections for balancing near the third critical speed are not quite in this diametral plane; certainly not the correction in plane 6. The latter could well relate to a mixed-mode condition which was definitely observed during modal balancing, as reported below.

Further confirmation of the presence of a fourth mode response at 3200 rpm is given by the phase angles of this residual whirl. The whirls at sensors 3 and 8 were in-phase with each other and out-of-phase to the motion at sensor 5, which strongly suggests a fourth mode whirl. It was noted that the balance masses, which had been attached to the shaft, had resulted in a very large mass in hole 20 in plane 4, together with fairly substantial masses in holes 7 and 9 at plane 4. These masses were replaced by a single, smaller mass in hole 20 at plane 4. Subsequent running of the shaft demonstrated that the effect of this change was to reduce slightly the whirl amplitude at the first critical speed and to leave the whirl at the third critical speed unchanged in amplitude.

## Balancing at the Fourth Critical Speed

Subsequently, resonant peaks in whirl amplitude were established at 7480 rpm and 7600 rpm; but at this stage in the balancing procedure, it was decided to perform a correction in planes 2, 4, and 6, using sensors 3, 5, and 8 at balancing speeds of 922, 3400, and 6800 rpm. Thus, the correction planes and measuring locations were the same as those selected for the two-speed balancing described above.

The initial whirl measurements made at these balancing speeds indicated fairly clearly that the significant residual unbalance at these speeds was primarily in the fourth mode. It was necessary to determine the appropriate influence coefficients, and this was performed by using a trial mass of 0.3475 grams, in turn, in planes 2, 4, and 6. For each of these planes, the resultant whirl was measured for the trial mass in hole 0 and again in hole 12 (i.e., at 0 and 180 degrees, respectively), as was done for the previous trial mass runs. The 0.3475 grams trial mass was the largest which could be used, such that the shaft could still be safely operated through its first and third critical speeds. Unfortunately, even this size trial correction did not have much effect on the whirl at 6800 rpm. The influence coefficients were calculated by vector subtraction of the relevant initial whirl vector from the subsequent whirl vector. From the initial whirl readings and the newly calculated coefficients, corrections were determined.

Despite the minimal change in the response of the fourth mode to the trial masses, these corrections were attached to the shaft. The correction masses were located in a large array of holes in order that the mass located at any one hole should not exceed 1.60 grams. Subsequent operation of the shaft showed that the whirl at the first and third critical speeds was increased substantially by these *correction* masses. In fact, the shaft could negotiate the first critical speed only by being rapidly accelerated through the critical speed; and the third critical speed was not negotiable at all, due to the large whirl.

These correction masses were left on the shaft; and a new two-speed trim correction was conducted at balancing speeds of 917 rpm and 3200 rpm, using the previous values for the relevant influence coefficients. The calculated corrections were fairly small. The effect of these corrections on the whirl was such that the amplitude of whirl at the first critical speed was reduced by about 50 percent, while at the third critical speed the reduction was of the order of 70 percent. Although the residual whirl levels were low enough to permit operation of the shaft through both critical speeds, the amplitudes were not as low as desired. The residual vibration readings were used with the old influence coefficients to calculate a second set of trim corrections. This set of corrections was attached to the shaft, and the residual vibration was further and significantly reduced at the first and third critical speeds. The test shaft was then run up to near the fourth critical speed region, and the amplitudes of vibration were observed to be much less than those present before the three-speed balance.

The next step was to return to a three-speed balance to further reduce the residual vibration levels for the fourth critical speed, while controlling the whirl at the first and third critical speeds. These vibration levels were used with the previously calculated, "three-speed" influence coefficients to calculate a trim correction. The effect of this latest set of corrections was to make the whirl somewhat worse near the first critical speed, but markedly better at 6800 rpm. The situation at 3400 rpm was not so clear cut. In fact, observation showed that the overall behavior was somewhat worse at the first and third critical speeds, but the whirl was much reduced at the fourth critical speed. In fact, it was fairly easy to negotiate the fourth critical speed.

At this point another set of corrections was calculated on the basis of the latest net whirl measurements, again using the previously determined influence coefficients. The corrections were applied to the shaft and left the whirl at 6800 rpm more or less unchanged, but produced improvements at 922 rpm and 3400 rpm. The residual whirl readings were once more used to calculate a further set of corrections which was subsequently installed. The whirl following this correction was such that there was slight improvement at the fourth critical speed, but no significant reduction of net whirl at the first and third critical speeds. It was still necessary to accelerate the shaft rapidly in order to safely traverse the third critical speed.

At this point, a two-speed trim correction, at 917 rpm and 3200 rpm, was calculated in the hope that the present satisfactory behavior at higher speeds would not be upset. The residual vibrations at these speeds were used, together with the earlier influence coefficients to predict some suitable corrections. These corrections were installed, and subsequent observation indicated a tremendous improvement in the shaft behavior at the first and third critical speeds and only a slight deterioration at the fourth critical speed. The shaft was in the best balanced condition which had been achieved by the influence coefficient technique, and no further corrections were attempted. A plot of the test shaft response at each of the major stages of the influence coefficient balancing tests is presented in Figure 8.7.

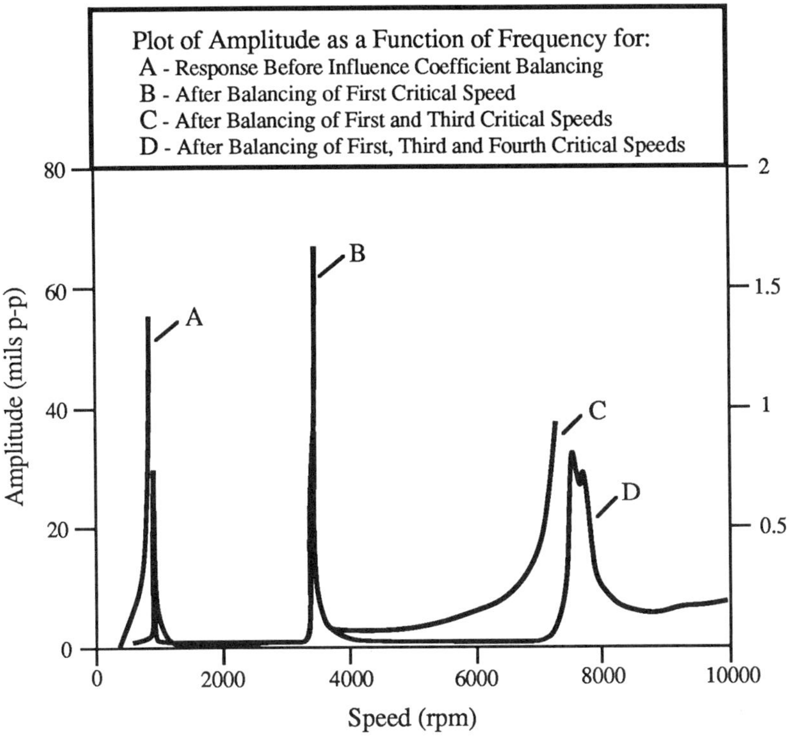


**Figure 8.7** Response of Supercritical Shaft Test Rig at Various Stages of the Influence Coefficient Balancing Tests

## Enhanced Modal Balancing Tests

Most of the correction masses which had been attached to the shaft during the influence coefficient tests were removed, so that the modal balancing experiments could be conducted with a test shaft which had a reasonable, but not gross, amount of unbalance. Although the tests reported in this section were concerned with a demonstration of modal balancing, advantage was taken of some of the on-line facilities which had been programmed for influence coefficient balancing. Thus, the shaft vibration at the selected measuring planes for the various balancing speeds was usually sampled five or ten times within close tolerances of the balancing speeds and the measurements automatically averaged. The vector subtraction implicit in determining the effect of a trial mass or masses was calculated using the relevant parts of the influence coefficient program, the necessary whirl measurements being provided to the computer on-line. The trial masses were also attached to the shaft at one angular location and then reattached at 180 degrees around the shaft. Both sets of associated whirl readings were then processed by the influence coefficient program to calibrate the shaft for the effects of the trial masses and reduce the errors inherent in data measurement. In addition, when more than one transducer was used along the shaft, the relevant results were subjected to the least-squares processing of the influence coefficient program in order to achieve a generally optimum rotor balance. Consequently, these tests were identified as *enhanced* modal balancing tests.

## Balancing the First Mode

A suitable balancing speed for the first mode correction was selected, at 901 rpm, and the vibration at this shaft speed was assumed to be largely in the first mode. No attempt was made to subtract out what seemed to be very small components of whirl from the higher modes. Only one transducer, sensor 5, was used in this part of the balancing process, and it was decided to add any necessary correction mass to the shaft in plane 4. A trial mass of 0.74 grams was attached to the shaft in plane 4 at an angular location of 120° and the shaft whirl again measured at sensor 5. The trial mass of 0.74 grams was then relocated around the shaft to 300° in plane 4, and the whirl at sensor 5 was again measured. These readings were then used to calculate the influence coefficient relating sensor 5 to plane 4 at 901 rpm. On this basis, a correction mass in plane 4 was calculated.

The residual net whirl of the shaft at 901 rpm showed that the whirl amplitude had been reduced by some 97 percent. The state of balance achieved was such that the shaft could be easily run through its first critical speed of 950 rpm and up to approximately 3400 rpm; that is, in the vicinity of the third critical speed. This was essentially identical to the procedure used during the influence coefficient tests. In fact, the on-line influence coefficient balancing computer program was used to calculate both the influence coefficients and the correction masses during the modal balancing of the first critical speed. As was noted above, the second mode response of the shaft was critically damped and, therefore, no balancing was required near the second critical speed.

## Balancing the Third Mode

As the second mode of the shaft was not significant in practice, modal theory suggested that correction masses located in two planes along the rotor should be sufficient to correct the third mode whirl, without upsetting the good behavior at the first critical speed. Moreover, in a simple modal interpretation at least, the two masses should be located in one diametral plane through the shaft.

That is, if the mass in one axial plane is fixed at an angular location of $q$ degrees, then the mass in the other axial plane should be located at $(q + 180)$ degrees around the shaft.

Considering the predicted third mode shape of the shaft, planes 2 and 6 were expected to be good axial locations for balancing the third mode, without disturbing the first mode correction. Consequently, prior to balancing the third mode, the sensitivity of the first mode to unbalance in planes 2 and 6 was determined experimentally for shaft speeds of 924 and 944 rpm, based on whirl measurements at sensor 5.

At 924 rpm, the response at sensor 5 to a trial mass in either plane 2 or 6 was virtually identical. Indeed the two influence coefficients were, in complex form:

$$a_{52} = 16.3 \text{ at } -8.5 \text{ degrees}$$
$$a_{56} = 16.3 \text{ at } -7.7 \text{ degrees}$$

By contrast, however, the situation at 944 rpm was more confused. The two influence coefficients were:

$$a_{52} = 74.9 \text{ at } -150.5 \text{ degrees}$$
$$a_{56} = 48.5 \text{ at } -146.1 \text{ degrees}$$

As 944 rpm is very close to the first critical speed, the influence coefficients at 944 rpm will be sensitive to slight variations in shaft speed and will be a much stronger function of shaft speed than the influence coefficients at 924 rpm. For this reason, reliance was placed on the data taken at 924 rpm, and it was assumed that the first mode would be insensitive to a suitable pair of equal masses, one in plane 2 at $q$ degrees, and the other in plane 6 at $(q + 180)$ degrees.

Sensors 3 and 8 were adopted as suitable measuring locations for balancing the third mode, and a balancing speed of 3200 rpm was selected. The initial net vibration of the shaft at this speed was measured. It is interesting to note that the difference in phase between the whirl at sensor 8 and that at sensor 3 at 3200 rpm was approximately 173 degrees, which is very close to the 180 degree value which might have been expected for an isolated third mode response. A trial set of masses, comprising 1.25 grams in plane 2 at 0 degrees and 1.25 grams in plane 6 at 180 degrees, was attached to the shaft and the whirl at sensors 3 and 8 again measured. It should be noted that the attachment of the trial set of masses produced no noticeable change in the shaft whirl at the first critical speed, confirming that the modal mass set was well chosen and orthogonal to the first mode. The trial set was then relocated around the shaft, so that in plane 2 the mass was at 180 degrees and in plane 6 the trial mass was at 0 degrees. The whirl of the shaft was again measured. From these observations, three calculations were made for suitable pairs of correction masses using the data averaging, error minimization and iteration features of the influence coefficient program. The calculated corrections were attached to the shaft and the residual whirl at 3200 rpm recorded. It was observed that, although the shaft behavior was not significantly changed at the first critical speed, there was a tremendous improvement in the third mode. Amplitude reductions at 3200 rpm at sensors 3 and 8 were of the order of 90 percent. Despite this marked reduction of whirl in the third mode, it was still not quite possible to negotiate the third critical speed at 3460 rpm.

Inspection of the whirl at 3200 rpm indicated that vibrations in planes 3 and 8 were now essentially in-phase, whereas before the addition of the last correction, the two vibrations had been almost exactly out-of-phase. The whirl at sensors 3 and 8 was also measured at 3400 rpm. The amplitudes were greater, being closer to the third critical speed, but the vibrations in planes 3 and 8 were once more out-of-phase. These observations suggested that a mixed-mode condition existed, such that the resonance aspect of the third mode was being partially masked by a significant whirl in the fourth mode. The presence of significant fourth mode response at 3200 rpm was also observed during the application of influence coefficient balancing to the third mode, as was indicated by the nature of one of the trim corrections. In addition, because the vibrations at sensors 3 and 8 at 3200 rpm had become in-phase after the addition of the pair of correction masses, the implications were that the presence of the fourth mode component had produced some over-correction of the third mode at 3200 rpm. A trim correction for the third mode was calculated from the whirl data at 3200 rpm and the modal *influence coefficients* deduced from the previous pair of trial masses. The trim correction was, to some extent, opposite in effect to the earlier corrections which reinforces the suspicion that the first pair of balance masses over-corrected the third mode at 3200 rpm.

Corrections masses were manufactured and attached to the shaft. The trim correction produced no obvious change in the shaft response at the first critical speed, but yielded a substantial reduction in the whirl amplitudes at the third critical speed, which could at this point be easily negotiated. More importantly, the residual whirl vectors at sensors 3 and 8 were both of roughly equal, small size and almost in-phase, thus confirming that the residual whirl is almost certainly due to unbalance in the fourth mode. Indeed it would have been possible to have balanced the two mixed modes simultaneously instead of adding a trim correction for the third mode. The two modes, however, were mixed in a simple manner, both modal defects having approximately the same orientation around the shaft. It was decided to correct only the residual defect in the third mode at 3200 rpm. The fourth mode unbalance was left to be balanced at a higher shaft speed, nearer the fourth critical speed, when its correction would be made more accurate by the resonant response in the fourth mode. The running record of shaft amplitude, which was maintained during the trials, indicated quite clearly the presence of some fourth mode response at the third critical speed of 3460 rpm.

It should be noted that the calculations of correction masses suitable for balancing the third mode by itself was made possible by the use of least-squares-minimization averaging of the data from the two sensors, a technique not generally associated with modal balancing. At this point, it was possible to easily and safely run the shaft through its first and third critical speeds. Consequently, no further attention was given to the first and third modes, and it was decided to proceed to balance the fourth mode of the shaft.

## Balancing the Fourth Mode

When the shaft speed was increased beyond the third critical speed, the whirl in the fourth mode was such that safe operation was not possible above 6400 rpm, even though the fourth critical speed was in the region of 7550 rpm. The vibration levels at 6400 rpm demonstrated the need to balance the shaft in its fourth mode. Consequently, 6400 rpm was selected to be the first balancing speed for the fourth mode, and planes 2, 4, and 6 were chosen for the attachment of correction masses. Again, since the second mode was critically damped, it could be ignored in the sense of modal balancing and only three balancing planes were necessary for balancing the fourth mode. Finally, it was decided to make vibration measurements at sensors 3, 5 and 8 and once more use some aspects of the influence coefficient program to optimize the predictions for correction masses deduced from the whirl measurements.

In balancing the fourth mode with a three-mass set in planes 2, 4 and 6, it is necessary to select an axial distribution of the three masses which is orthogonal to the first and third modes of the shaft. In this way, any masses used to correct the fourth mode of unbalance will not upset the previously attained balance in the first and third modes. For this purpose, it was necessary to determine the modal sensitivities of masses in planes 2, 4 and 6 in the first and third modes. In fact, it will be remembered that tests preliminary to the third mode correction had already established that a mass had approximately the same effect on the first mode, whether it was attached to the shaft in plane 2 or plane 6.

Some of these modal sensitivity tests were performed at a convenient time while a set of correction masses was being manufactured for balancing of the third mode. The tests were conducted at 917 rpm and 3200 rpm using a trial mass of 1.25 grams. The trial mass was located successively in planes 2, 4 and 6 and at angular positions of 0 degrees and 180 degrees. In each case, the subsequent whirl of the shaft was compared with the corresponding value with no trial mass attached. Subtraction of corresponding vibration vectors gave the effect of the trial mass alone, and from this a modal influence coefficient was effectively calculated.

Two inconsistencies were found in the results of these calculations. First, whereas earlier tests at 917 rpm had suggested that a mass in plane 2 had the same effect as an equal mass in plane 6, the present test indicated a difference in sensitivity between planes 2 and 6 of approximately 9 percent; although, of course, this was still a fairly small difference. The results for the third critical speed at 3200 rpm were more puzzling. A mass added at plane 2 seemed to produce a greater response at sensor 8 than at sensor 3. Conversely, a mass in plane 6 apparently resulted in a greater whirl at sensor 3 compared with sensor 8. Remembering that plane 2 and sensor 3 are close together, as are plane 6 and sensor 8, these results were unexpected. It had been thought that only modes 3 and 4 were significant at 3200 rpm. The above behavior, however, suggested that there was also some response in the first mode at 3200 rpm. This is apparent from the nature of the first, third and fourth mode shapes and the differing phase relationships between the unbalance force produced by the trial mass and the corresponding response of these modes; the speed of 3200 rpm being greater than the first critical speed, but less than the third and fourth critical speeds. That is, response to unbalance in the first mode would be approximately 180 degrees out-of-phase with the unbalance mass, while for the other modes the response would be approximately in-phase with the unbalance mass.

Overall, this set of sensitivity tests suggested that a suitable distribution of correction masses for the fourth mode would be a mass $m$ at some angle $q$ degrees in plane 2, a mass $1.4m$ at an angle $(q + 180)$ degrees in plane 4, and a mass $m$ at $q$ degrees in plane 6. Balancing the fourth mode then requires making the correct choice of $m$ and $q$.

The shaft was run up to 6400 rpm and the whirl at sensors 3, 5 and 8 was measured. It was noted that the net whirl at sensors 3 and 8 was almost exactly in-phase, and both were approximately out-of-phase with the motion at sensor 5, thus confirming that the motion was predominantly in the fourth mode. Next, a trial mass distribution of 1 gram at 0 degrees in plane 2, 1.4 grams at 180 degrees in plane 4, and 1 gram at 0 degrees in plane 6 was attached to the shaft, and the whirl at sensors 3, 5 and 8 at 6400 rpm was again measured. The trial mass distribution was then moved around the shaft, so that it became 1 gram at 180 degrees in plane 2, 1.4 grams at 0 degrees in plane 4, and 1 gram at 180 degrees in plane 6. The whirl of the shaft was again measured at sensors 3, 5 and 8 at 6400 rpm. From these results, a suitable balance distribution for the fourth mode was calculated, using the data-averaging and error-minimization features of the influence coefficient program. The program, in fact, suggested three distinct mass distributions through iteration, but all three were very similar which implied consistency and encouraged confidence in the results.

Interestingly, these corrections were to be attached to the shaft in approximately the same diametral plane through the shaft as were the corrections for the third critical speed. Moreover, both corrections have the same phase as the whirl of the shaft at 3200 rpm and 3400 rpm during third mode balancing. This confirms the mixed-mode condition hypothesized while balancing the third mode, as to the presence of defects in the third and fourth mode, both of which caused significant whirl components.

This correction was attached to the shaft and the shaft was run up in speed. Observation showed that the amplitude of the shaft was slightly increased at the first critical speed of 950 rpm, but the first critical speed was still easily negotiated. This slight increase was due, presumably, to the approximate nature of the modal proportions of 1 : −1.4 : 1 for the three correction masses. The behavior of the shaft was unchanged at the third critical speed. It was possible to run the shaft up to 10,000 rpm, right through the fourth critical speed region, with no unacceptable levels of vibration. There were two resonant peaks at 7480 rpm and 7600 rpm, but both were within satisfactory limits. Indeed, at the balancing speed of 6400 rpm, this correction reduced the amplitudes by better than 90 percent. Consequently, the shaft was deemed to be balanced satisfactorily for operation up to 10,000 rpm.

Although the shaft was now satisfactorily balanced, two further investigations were carried out. First, a trim correction set of masses in planes 2, 4 and 6 was calculated for the residual whirl, using the previously established effect of the three modal trial masses at 6400 rpm. The suggested trim masses were less than 0.3 grams in planes 2 and 6 and about 0.4 grams in plane 4. As these masses are very much smaller than the initial corrections of 8.5 grams in planes 2 and 6 and 11.89 grams in plane 4, the trim correction set was not used.

Finally, it was decided instead to use a single trim correction in plane 4 to reduce the amplitude at the first critical speed and possibly at the fourth critical speed. This mass, of course, was not likely to change the shaft behavior at the third critical speed, as plane 4 is located near a node for the third mode. The shaft whirl was measured at 900 rpm using a sensor in plane 5 and correction of 0.80 grams at 160.9 degrees in plane 4 was calculated. The shaft was run with this final correction attached, and the result was very successful. There was a great reduction in amplitude at the first critical speed, no change at the third critical speed, and a further slight improvement in shaft behavior in the region of the fourth critical speed. At this stage, the test shaft was in the best balanced condition throughout the speed range up to 10,000 rpm that had been achieved by either modal or influence coefficient balancing. A plot of the test shaft response of each of the major stages of the modal balancing is presented in Figure 8.8.

It was found that the angular locations of the corrections added during modal balancing were very similar to the locations from which the previous masses were removed. Indeed, the modal correction in plane 6 was virtually the same as the mass which was removed, but the magnitudes of the masses added in planes 2 and 4 were not in close agreement with the masses removed. There are at least two reasons for this discrepancy. First, modal balancing did not exactly reproduce the vibration state of the rotor at the completion of influence coefficient balancing. Second, the corrections in modal balancing may have been redistributed in part between planes 2 and 4. It must be stressed that the data presented here for the mass removed from the shaft before starting modal balancing was not aggregated in this form until the tests were completed. The information was not available during the modal balancing tests.

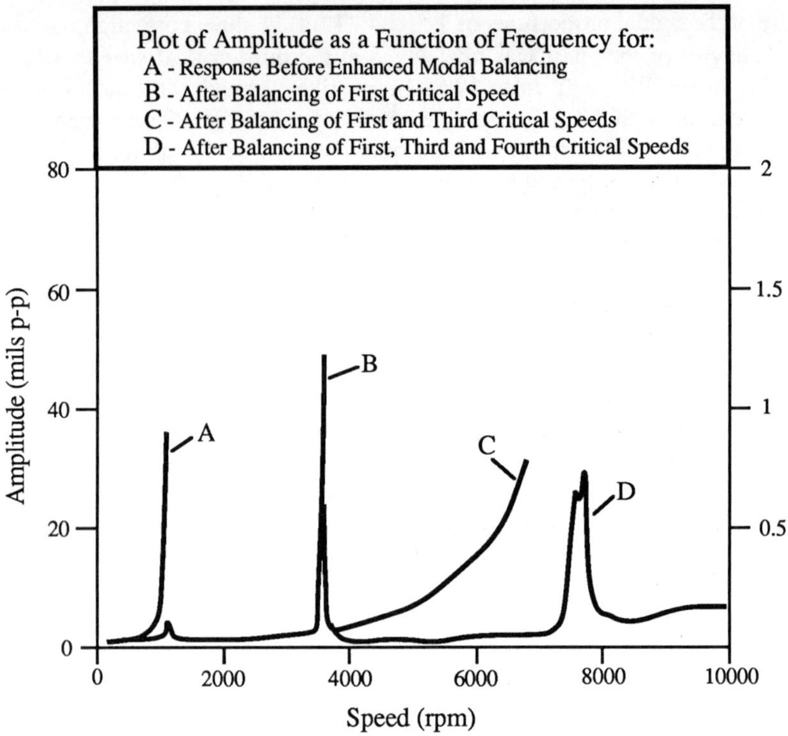


**Figure 8.8** Response of Supercritical Shaft Test Rig at Various Stages of the Enhanced Modal Balancing Tests

## Unified Balancing Approach Tests

The Unified Balancing Approach was used for the third set of balancing tests. These tests used an on-line computer program similar to that used for the first two sets of balancing tests. However, the software was reconfigured for the procedure and the computations of the Unified Balancing Approach. Details of these tests follow [35].

## Balancing at the First Critical Speed.

At the start of this set of tests, the condition of the rotor was such that the first critical speed could not be negotiated. In fact, the test rig could not be operated safely above 900 rpm. This initial condition was comparable to the unbalance condition prior to each of the other test programs. Since a large correction was anticipated for the first critical speed, the correction was spread to three balance planes, numbers 2, 4 and 6, by installing 3 trial masses simultaneously. Note that the procedure for balancing at the first critical speed was effectively identical for all three balancing methods, as applied in these tests.

Uncorrected rotor data and modal trial data, for equal masses in the three planes, were taken for the first mode at 890 rpm. A correction mass of about 2.3 grams at about 330 degrees in each of planes 2,4 and 6 was calculated. These masses were fabricated and installed. The test rig was run and check balance data were taken. The net synchronous vibration of the rotor was reduced by more than 40 percent, but the first critical speed was not yet negotiable. A trim correction of about 1.3 grams at about 270 degrees in each of planes 2, 4 and 6 was calculated. These correction masses were fabricated and installed. The response at the first critical speed was reduced tremendously and the first critical speed was negotiated easily. Check balance data, which indicated the presence of measurable residual vibration, were then taken. Based on these data, a set of trim corrections of about 0.6 grams at 165 degrees in each of planes 2, 4 and 6 was calculated. These correction masses were subsequently fabricated and installed

With the third set of correction masses installed, the test rig was run and check balance data was taken. These data indicated a substantial improvement; better than a 95 percent reduction of the synchronous vibration as compared to the original unbalanced condition. The vibration at the first critical speed was well controlled and the test rig was capable of safe operation up to about 7800 rpm. With the change in the damper design, the fourth critical speed was increased to about 8000 rpm. With the new damper design, the second critical speed, at about 2500 rpm, was no longer critically damped. However, it was heavily damped and relatively unresponsive to unbalance. Therefore, there was no point in conducting any balancing tests at the second critical speed. The third critical speed, on the other hand, although well damped, did evidence some response to unbalance. Thus, the balancing procedure was continued with the third critical speed, treated as the second mode.

## Balancing at the Third Critical Speed

During the individual trial mass portion of the balancing procedure for the third critical speed, planes 2 and 4 were mistakenly used, instead of planes 2 and 6 as originally intended. Thus, the modal trial mass set that was calculated was not longitudinally symmetric, as it was expected to be for planes 2 and 6. That is, it was anticipated that the modal trial set for planes 2 and 6 would be composed of a mass in plane 2 approximately equal in magnitude and opposite in direction to that in plane 6. However, the use of planes 2 and 4 resulted in a modal trial mass set composed of a mass in plane 2 approximately 50 percent larger in magnitude than that in plane 4, with the angles of the two masses differing by about 170 degrees. Clearly, the combination of planes 2 and 4 was not the optimum plane combination to be used for balancing the third critical speed. However, it was anticipated that the use of such non-ideal planes might provide an interesting test of the Unified Balancing Approach. Consequently, the modal trial mass set composed of planes 2 and 4 was used for balancing the third critical speed. The modal correction mass set that was calculated and installed resulted in better than a 70 percent reduction of the response at the third critical speed, while having no discernible effect on the response at the first critical speed. A trim correction was calculated, but it was not installed because it was quite small and the response at the third critical speed was already well controlled.

The test rig was not run again until the following day. Without adding or removing any trial or correction masses, the test rig was run up to a speed of about 7800 rpm. Although the synchronous response at the second and third critical speeds was not noticeably changed from the previous run, the response at the first critical speed was considerably degraded. Apparently, something in the shaft or test rig had changed spontaneously to alter the amount or distribution of unbalance, or the amount of damping. It is possible that the change in the test rig behavior was due to creep in the shaft or in the elastomer damper elements. In any case, the balancing procedure was continued with a trim balance at the first critical speed.

For this trim balance operation, a data speed of 911 rpm was used. Since this speed was closer to the first critical speed than the balancing speed of 890 rpm used during the earlier runs, more sensitive balancing data could be acquired. The uncorrected rotor data taken during this balancing run indicated that the synchronous response of the test rig near the first critical speed was more than double what it had been during the previous run. The trial mass, and subsequently the correction mass, was composed of just a single mass in plane 4, rather than being distributed between planes 2, 4 and 6, as was done previously. This change was made for two reasons: first, since a much smaller correction mass was anticipated; and, second, to minimize the effect of such a correction on the response at the third critical speed as plane 4 is located very near a node in the corresponding mode shape. As a result of this balancing run, a correction mass of 1.7 grams at about 44 degrees was calculated and installed in plane 4. A set of check balance data for the first mode was taken and a reduction in synchronous response, as measured by the various probes, of between 75 and 93 percent was indicated. As expected, this correction mass had no discernible effect on the response at the third critical speed.

It was not necessary at this point to do any more balancing at any of the first three critical speeds. Before continuing balancing with the fourth critical speed, several test runs were conducted and the response of the test rig was observed, particularly in the vicinity of the first critical speed, to evaluate the repeatability of the response. The response was very repeatable except around the first critical speed where some substantial variations in response were observed. These variations were generally within 25 percent of the *balanced* response, which is actually fairly small in absolute magnitude. However, at one point the magnitude of the response at the first critical speed increased by more than a factor of two, only to return to its *balanced* response on the following run. There appeared to be a shift in the test shaft followed by a shift back. After this run, the test rig response became more repeatable, so the balancing tests were resumed with the balancing of the fourth critical speed.

## Balancing at the Fourth Critical Speed

Uncorrected rotor data were taken for the first three modes (e.g. at the first, third and fourth critical speeds), followed by a series of individual trial mass runs for planes 2, 4 and 6. Data for these trial mass runs were taken in the vicinity of the first and third critical speeds to permit the calculation of a modal trial mass set for the fourth critical speed, treated as the third mode. The resulting modal mass set consisted of a mass of $p$ grams at $q$ degrees in plane 2, which was treated as the reference plane, a mass of $1.4p$ grams at $q+191$ degrees in plane 4 and a mass of $1.17p$ grams at $q+3.6$ degrees in plane 6. Note that this modal mass set was almost, but not quite, planar. A balancing run was conducted for the fourth critical speed using a modal trial mass set, based on this modal mass set, with $p$ chosen to be unity and $q$ chosen to be zero. The modal trial mass set had no noticeable effect on the response at any of the first three critical speeds, neglecting the run-to-run variation at the first critical speed. A modal correction mass set was calculated for the fourth critical speed. This set consisted of a mass of about 1.3 grams at 30 degrees in plane 2, 1.8 grams at 221 degrees in plane 4 and 1.5 grams at 34 degrees in plane 6. These corrections were fabricated and installed. The fourth critical speed was subsequently easily negotiated. There was no discernible effect on the response at either the second or third critical speed, and any change in the response at the first critical speed could easily be attributed to run-to-run variation in that response. The check balance data taken at the balancing speed for the fourth critical speed, 7620 rpm, indicated a reduction in synchronous rotor response of better than 90 percent. A trim balance mass set was calculated, but the masses were impractically small and the rotor response was so good that the trim correction was not installed. This established a new, best ever operating condition for this test rig.

The modal mass set ratios for the Unified Balancing Approach tests are listed in Table 8.1 and are illustrated schematically in Figures 8.9 and 8.10, where only the plane 4 mass is shown for the first mode. The axial distribution for the modal mass sets is shown in Figure 8.9; the angular distribution is shown in Figure 8.10. Note that the sets are almost, but not quite, planar. The sequence of balancing operations in this set of tests is illustrated in Figure 8.11 by a series of vector plots, pairs of which show the correction mass added in the reference plane for each mode and the resulting change in vibration for the corresponding mode. A set of typical rotor response plots is presented in Figure 8.12.

**Table 8.1** Modal Mass Set Ratios From Unified Balancing Approach Tests

| UBA Mode No. | Plane 2 | Plane 4 | Plane 6 |
|---|---|---|---|
| 1 | --- | 1.0 at 0° | --- |
| 2 | 1.0 at 0° | 0.67 at −171° | --- |
| 3 | 1.0 at 0° | 1.39 at −169° | 1.17 at 3.6° |

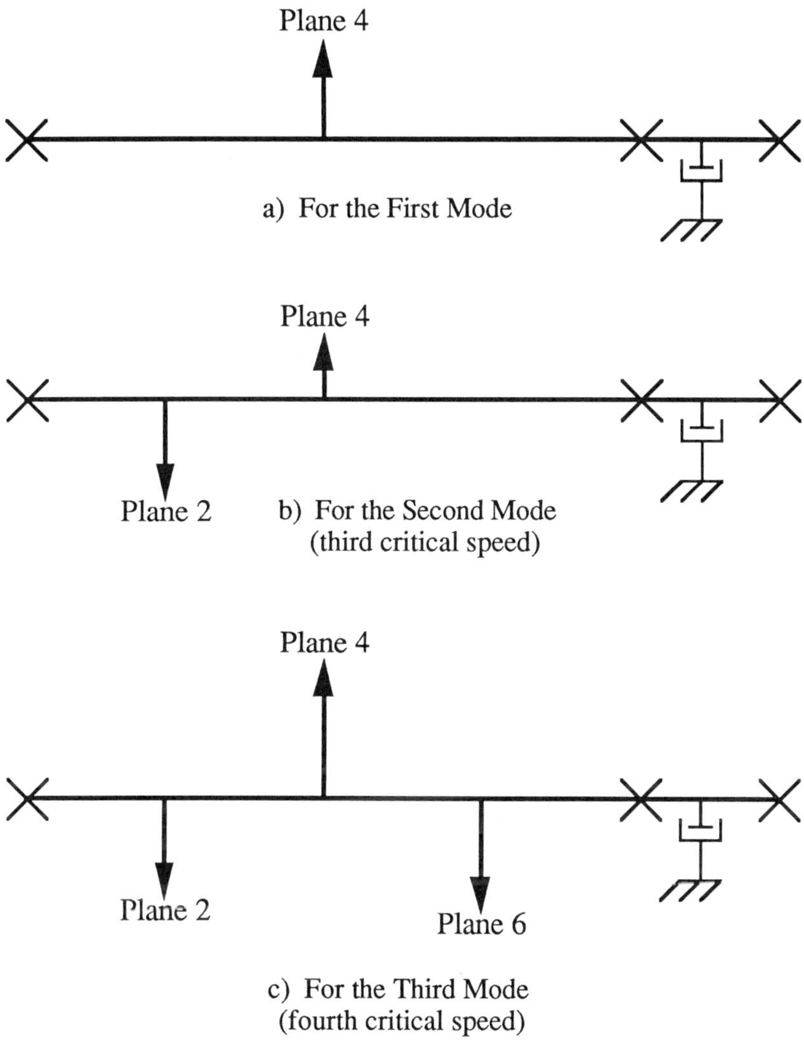


**Figure 8.9** Longitudinal Distribution of Modal Mass Sets

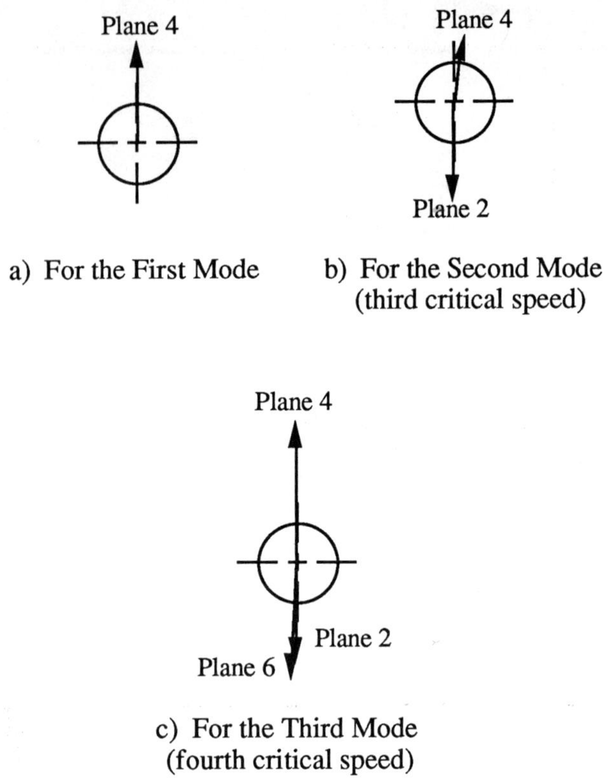


**Figure 8.10** Angular Distribution of Modal Mass Sets

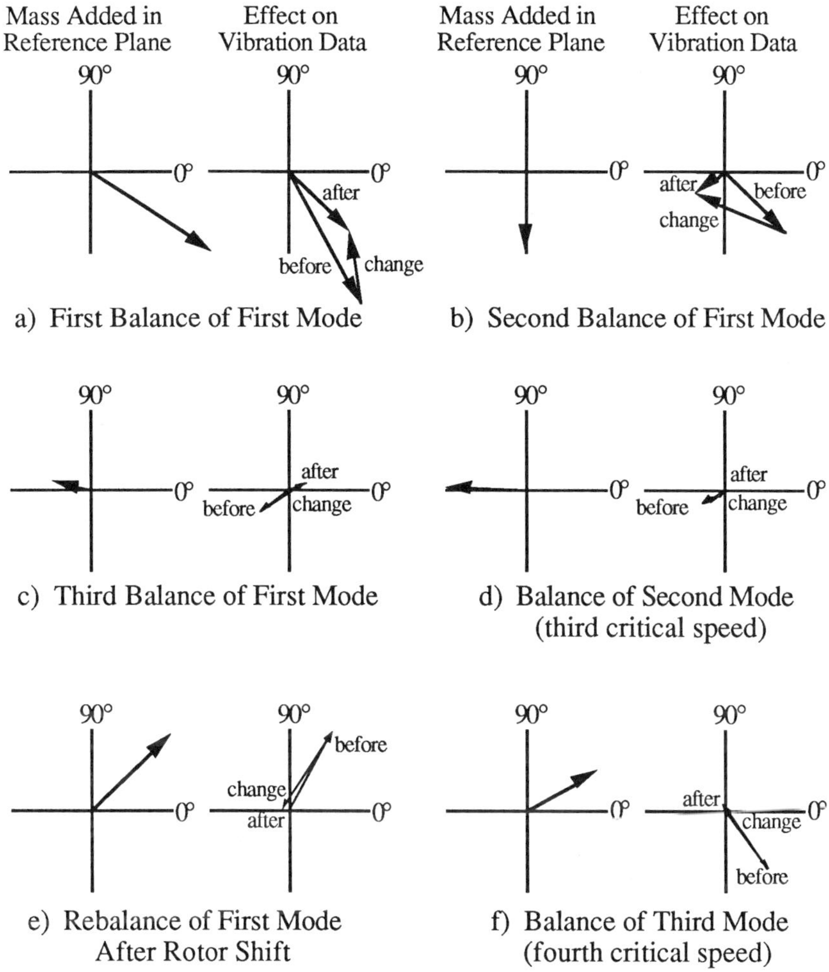


**Figure 8.11** Effect of Correction Masses on Vibration Data for Unified Balancing Approach Tests

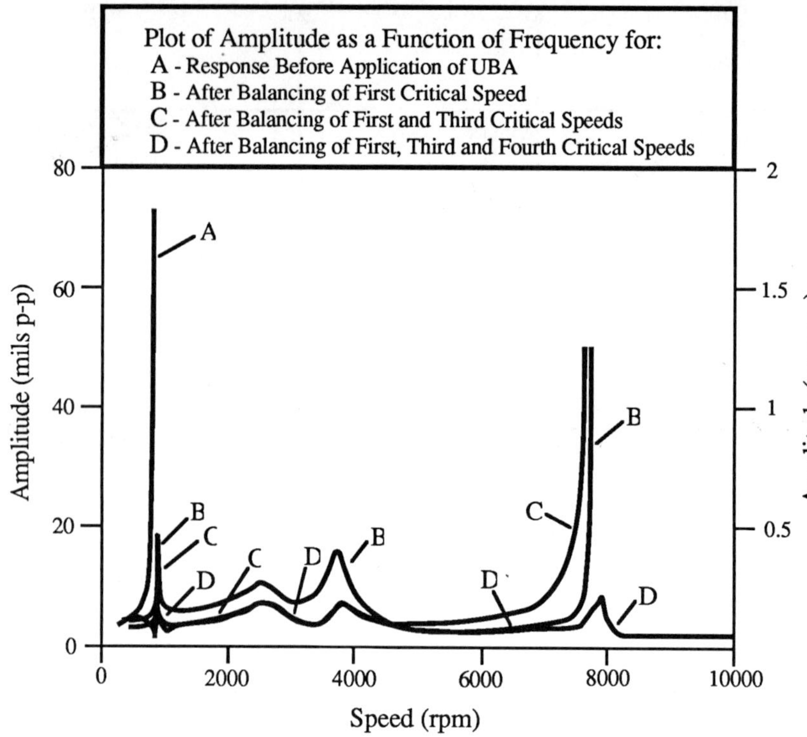


**Figure 8.12** Response of Supercritical Shaft Test Rig at Various Stages of the Unified Balancing Approach Tests

# Chapter 9
# Application of the Principle of Reciprocity to Flexible Rotor Balancing

The least desirable feature of most flexible rotor balancing procedures is the considerable number of trial mass runs required. This is of particular importance in the balancing of machines which require a substantial stabilization time during start-up. Using an adaptation of the principle of reciprocity, it is possible to significantly reduce the required number of trial mass runs for certain rotors when using either influence coefficient balancing or the Unified Balancing Approach. This may be applied to the more empirical forms of modal balancing, as well. When applied to flexible rotor balancing, the principle of reciprocity states that, given two rotor axial locations, $A$ and $B$, at which both balancing planes and vibration sensors are located, the influence coefficient relating the vibration level at $A$ to the unbalance at $B$ is identical to that relating the vibration level at $B$ to the unbalance at $A$. This is true even in the presence of damping. This chapter begins with a theoretical discussion of the principle of reciprocity and its application to flexible rotor balancing. The particular means by which reciprocity can be applied to improve the influence coefficient and Unified Balancing Approach procedures are then described in detail. Results of a numerical study to verify this application of reciprocity and to investigate any possible limitations are also discussed, along with the results of a similar experimental study using two substantially different tests rotors.

## Principle of Reciprocity

The principle of reciprocity was first proposed by Maxwell [89]. He defined it only as it related to static structures. That is, given a unit force applied at point $A$ in a structure and the resultant deflection at point $B$; applying a unit force at point $B$ would then result in an identical deflection at point $A$. This could, of course, be related as well to applied deflections and corresponding resultant forces.

Some years later, Rayleigh [111] applied this principle to dynamic loading of structures. He showed analytically that the principle holds true in the presence of damping, and that the damping need not be viscous. Thus, the foregoing definition was extended to the case where the forces and deflections oscillate. In this case, the deflections are identical in terms of both their amplitudes and their phase relationships to their respective forces.

In the years since, the application of the principle of reciprocity to the unbalance response of rotating shafts has been considered by Kroon [74], and possibly others. Darlow and Smalley [31] subsequently investigated a means to take advantage of reciprocity in rotor balancing. They indicated that reciprocity is apparently of little or no benefit to rigid rotor balancing or to nonempirical modal balancing of flexible rotors. Also, while the benefits to influence coefficient balancing of flexible rotors can be substantial, these benefits are restricted to rather limited situations. However, the benefits to the Unified Balancing Approach and to empirical modal balancing methods are at least as dramatic as for influence coefficient balancing, while not nearly as limited. The means by which this is accomplished are discussed in this chapter. While this chapter is concerned primarily with application to the Unified Balancing Approach, empirical modal balancing methods can be enhanced in a similar manner. Also, for any balancing method, reciprocity may be used to check the reliability of balancing data.

It is also important to establish that the principle of reciprocity does in fact apply to real rotordynamic systems and will also apply in the presence of damping, gyroscopic effects, and cross-coupling bearing effects. To this end, a series of numerical and experimental studies were conducted by Darlow and Smalley [31] and the results are reproduced in this chapter.

When determining the validity of reciprocity for rotordynamic systems, it is important to consider certain influences or parameters which were not included in Rayleigh's analysis or are peculiar to rotating systems. The most important of these are gyroscopic effects, cross-coupling bearing parameters, and system nonlinearities. The effect of nonlinearities can be essentially neglected because of the nature of rotor balancing procedures. That is, since virtually all rotor balancing procedures depend on linear behavior, balancing must progress in sufficiently small steps for the system to appear linear. Thus, if the system appears linear as regards balancing, it will most certainly also appear linear as regards reciprocity. In contrast, the influence of gyroscopic and cross-coupling effects cannot be ignored. Thus, a numerical study to consider these effects is discussed in a later section.

As applied to rotor balancing, reciprocity is used to generate additional influence coefficients without additional trial mass runs, and thus to reduce the total number of runs required. The specifics of the application of reciprocity and the benefits gained are also discussed in a later section. In order for reciprocity to be applied to rotor balancing, it is necessary that displacement probes be located adjacent to each of the balance mass planes. In general, the axial distance between the planes and adjacent probes is less critical when balancing the lower modes and more critical when balancing the higher modes. It is also important to take care that the trial mass location associated with a particular mode is not located too near a nodal point for that mode.

## Numerical Verification of Reciprocity

A numerical investigation was conducted to verify the principle of reciprocity for the unbalance response of rotordynamic systems using a transfer matrix analysis. Two different rotor models were used. The first model is a direct representation of one of the test rigs used in the experimental investigation described in the next section. The second is a more general model designed to take into account gyroscopic and cross-coupling effects, independently and in combination. In both cases, influence coefficient matrices were generated for a particular speed and set of axial locations. Reciprocity is indicted by symmetry about the main diagonal of the influence coefficient matrix. The elements of the matrices are complex.

The first rotor model, referred to below as the *demo rig*, is illustrated by the sketch in Figure 9.1. For this model, the unbalance response was calculated at each of three speeds in the vicinity of the first critical speed for small unbalance at each of four axial locations. The results of this analysis, as indicated by the symmetry of the influence coefficient matrices presented in Table 9.1, are in complete support of the principle of reciprocity. Since this model had very little damping (rigidly mounted ball bearings) and was not run right at the critical speed, all of the influence coefficients had angles very near to either zero or 180 degrees. Thus only the magnitudes of the influence coefficients are given in Table 9.1, where the negative signs indicate angles of approximately 180 degrees.

A sketch of the second rotor model is presented in Figure 9.2. The rotor has a length of 3.42 m (134.5 in.). Table 9.2 shows the bearing coefficients used.

Four cases were investigated using this analytical model:

1. No rotary (gyroscopic) inertia; no cross-coupling stiffness at bearings
2. Rotary inertia; no cross-coupling stiffness at bearings
3. No rotary inertia; cross-coupling stiffness at bearings
4. Rotary inertia; cross-coupling stiffness at bearings

The different cases had notably different peak amplitudes and the speed for the peak amplitude varied among 3100 rpm for Case 1, 3200 rpm for Case 2, and 3400 rpm for Cases 3 and 4. Stations 8 and 12 were used to investigate reciprocity at 1.71 m (67.5 in.) and 2.77 m (108.875 in.) from station 1, respectively. Station 12 is 0.65 m (25.625 in.) from the last station in the rotor. Thus, the two stations are by no means symmetrically located on the rotor.

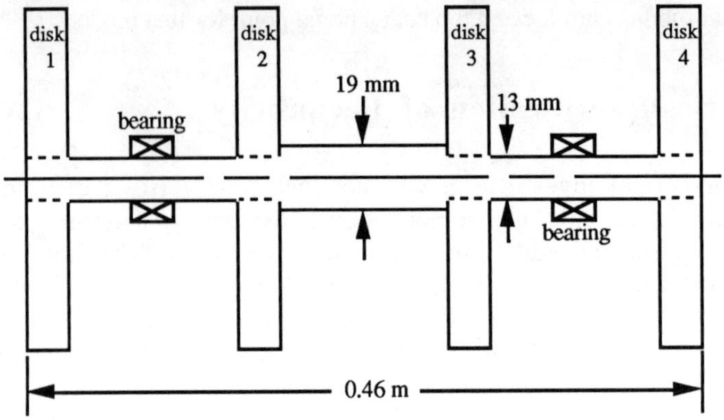


**Figure 9.1** Demo Rig for Reciprocity Tests

**Table 9.1** Numerically Determined Influence Coefficients for Demo Rig (mm/gr-mm x $10^4$)

| Speed (rpm) | Sensor No. | Balancing Plane No. 1 | 2 | 3 | 4 |
|---|---|---|---|---|---|
| 4600 | 1 | 8.68 | -3.77 | -3.38 | 5.43 |
| 4600 | 2 | -3.77 | 2.37 | 2.11 | -3.38 |
| 4600 | 3 | -3.38 | 2.11 | 2.37 | -3.77 |
| 4600 | 4 | 5.43 | -3.38 | -3.77 | 8.68 |
| 4800 | 1 | 13.4 | -6.10 | -5.64 | 9.59 |
| 4800 | 2 | -6.10 | 3.60 | 3.33 | -5.64 |
| 4800 | 3 | -5.64 | 3.33 | 3.60 | -6.10 |
| 4800 | 4 | 9.59 | -5.64 | -6.10 | 13.4 |
| 5000 | 1 | 28.4 | -13.7 | -13.2 | 23.9 |
| 5000 | 2 | -13.7 | 7.58 | 7.27 | -13.2 |
| 5000 | 3 | -13.2 | 7.27 | 7.58 | -13.7 |
| 5000 | 4 | 23.9 | -13.2 | -13.7 | 28.4 |

(Note: positive values indicate a phase angle of approximately zero degrees and negative values indicate a phase angle of approximately 180 degrees)

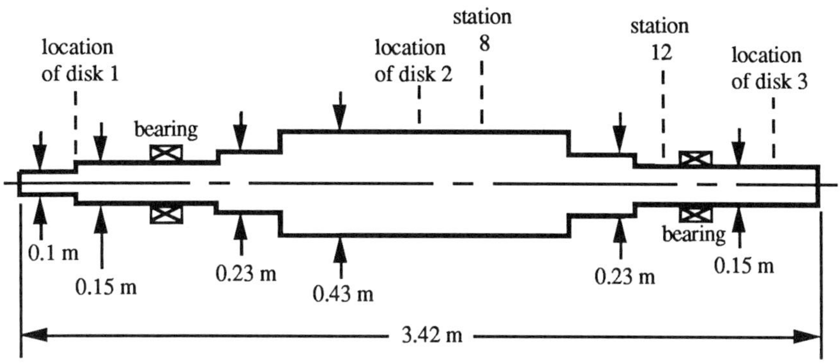


**Figure 9.2** Second Rotor Model for Numerical Reciprocity Study

**Table 9.2** Bearing Stiffness and Damping Coefficients for Second Rotor Model (N/m)

| | Cases 1 and 2 | Cases 3 and 4 |
|---|---|---|
| $k_{xx}$ | 6.30 x $10^8$ | 6.30 x $10^8$ |
| $k_{xy}$ | 0 | 13.50 x $10^8$ |
| $k_{yx}$ | 0 | 3.30 x $10^8$ |
| $k_{yy}$ | 5.60 x $10^8$ | 5.60 x $10^8$ |
| $\omega B_{xx}$ | 13.00 x $10^6$ | 13.00 x $10^6$ |
| $\omega B_{xy}$ | 0 | 0 |
| $\omega B_{yx}$ | 0 | 0 |
| $\omega B_{yy}$ | 4.55 x $10^6$ | 4.55 x $10^6$ |

Table 9.3 presents the results for the four cases as influence coefficient matrices relating response at stations 8 and 12 to unbalance at stations 8 and 12. Symmetry may be measured by how closely the off-diagonal elements are in agreement. In each case, the influence coefficient matrix is presented for the speed of greatest response. The symmetry is reasonable for all cases. The angles of the off-diagonal elements agree within 0.1 degree in all cases and the discrepancy between the corresponding magnitudes varies between 0.5 percent (Case 1) and 3 percent (Case 4). It is recognized that a 3 percent discrepancy is non-trivial, but its magnitude is small compared to the factor of 6 variation in response between Cases 1 and 4. Also, a 3 percent discrepancy is not large

when considering the measurement error generally associated with practical rotor balancing data (typically 5-10 percent).

**Table 9.3** Numerically Determined Influence Coefficients for Second Rotor Model (mm/gr-mm x $10^4$)

| Case No. | Gyro? | Cross-Coupling? | Influence Coefficient Matrix | | % Assym. |
|---|---|---|---|---|---|
| 1 | No | No | 25.7 at -14.4° | 12.9 at -14.7° | 0.5 |
| | | | 12.8 at -14.8° | 6.67 at -14.5° | |
| 2 | Yes | No | 37.7 at -22.8° | 19.3 at -23.2° | 2.6 |
| | | | 18.8 at -23.2° | 9.91 at -22.9° | |
| 3 | No | Yes | 6.46 at 93.2° | 2.70 at 100.7° | 0.3 |
| | | | 2.69 at 100.8° | 1.28 at 96.8° | |
| 4 | Yes | Yes | 5.29 at 108.4° | 2.25 at 117.2° | 3.0 |
| | | | 2.32 at 117.2° | 1.10 at 108.6° | |

## Experimental Verification of Reciprocity

Two substantially different test rigs were used to experimentally verify the principle of reciprocity for the unbalance response of rotordynamic systems. One of the test rigs, the supercritical shaft test rig described in Chapter 8, had an essentially uniform shaft. The other test rig, the *demo rig*, had a slender shaft on which were mounted four relatively massive disks.

## Supercritical Shaft Verification

The supercritical shaft test rig was a model of an advanced helicopter power transmission shaft described in detail in Chapter 8 and shown in Figure 9.3. The influence coefficient matrices for several speeds for the supercritical shaft test rig are presented in Table 9.4. These data were retrieved from the balancing tests described in Chapter 8, covering the first three modes for this test rig. Most off-diagonal coefficients exhibit good symmetry. The least amount of symmetry occurs in influence coefficients relating to the shaft center for the second mode, at 3400 rpm, corresponding to a nodal point. Even in these cases, the divergence from symmetry is small in absolute magnitude and may be attributed to the axial separation of several inches between balancing planes and vibration sensors.

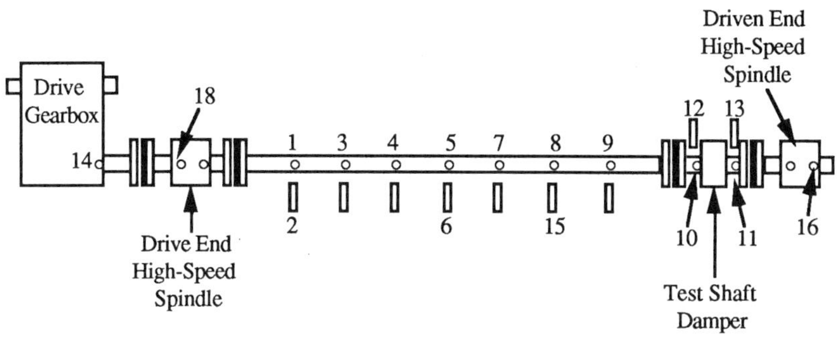


**Figure 9.3** Supercritical Shaft Test Rig

**Table 9.4** Measured Influence Coefficients for Supercritical Shaft Test Rig

| Speed (rpm) | Sensor No. | Balancing Plane No. 2 | 4 | 6 |
|---|---|---|---|---|
| 926 | 3 | 14.8 at 346° | 19.4 at 353° | 14.2 at 351° |
| 926 | 5 | 20.8 at 342° | 28.9 at 353° | 20.9 at 344° |
| 926 | 8 | 13.3 at 345° | 19.7 at 348° | 16.9 at 348° |
| 3400 | 3 | 10.4 at 348° | 0.96 at 274° | 10.6 at 171° |
| 3400 | 5 | 0.70 at 220° | 0.90 at 184° | 1.31 at 166° |
| 3400 | 8 | 10.8 at 188° | 0.98 at 93° | 9.61 at 342° |
| 6800 | 3 | 0.21 at 40° | 3.11 at 178° | 2.38 at 357° |
| 6800 | 5 | 3.24 at 183° | 2.56 at 2° | 2.58 at 187° |
| 6800 | 8 | 1.86 at 13° | 2.60 at 182° | 0.40 at 187° |

## Demo Rig Measurements

A sketch of the demo rig is shown in Figure 9.1. This rig is composed primarily of a slender shaft about 46 cm (18 in.) long, mounted on ball-bearings, with four relatively massive disks; two straddle-mounted and two overhung.

In this case, influence coefficients were determined using the linear regression method developed by Larsson [77]. A series of trial mass runs were performed and the influence coefficient set was determined, based on a minimization of the sum of the squared errors between measured response to the trial masses and predicted response using the influence coefficients. Figure 9.4 shows how the influence coefficients developed or converged as the number of trial mass runs on which they were based increased. After initially scattered results for the minimum number of trials, the calculated influence coefficients converged to values symmetrical to within 9 degrees in phase angle and to within 1.6 percent in magnitude.

## Using Reciprocity in Flexible Rotor Balancing

Often when using influence coefficient balancing or the Unified Balancing Approach for balancing flexible rotors, the total number of required runs can be reduced by utilizing the principle of reciprocity to infer some influence coefficients when their symmetrical counterparts are known. Reciprocity may alternatively be used to check the reliability of balancing data.

To use reciprocity with influence coefficient balancing, a single trial mass is used for one or more data speeds and influence coefficients are determined for a number of probes for each speed. The influence coefficient matrix is then expanded by utilizing symmetry. Reciprocity will be useful in influence coefficient balancing only when balancing more than one speed, and will be of significant benefit only if the number of balancing planes required to get good sensitivity at all speeds is substantially less than the number of speeds. Clearly, the greatest benefit will result if a single balancing plane can be found which can achieve good sensitivity at all balancing speeds. For example, ordinary influence coefficient balancing of four critical speeds, one at a time, using one plane for the first critical speed and going up to four planes for the fourth critical speed, will require a minimum of ten trial mass runs. If a single balancing plane can be found which is effective at all critical speeds, the minimum number of trial mass runs can be reduced to four through the use of reciprocity. This ideal example represents a 60 percent reduction in the minimum number of trial mass runs.

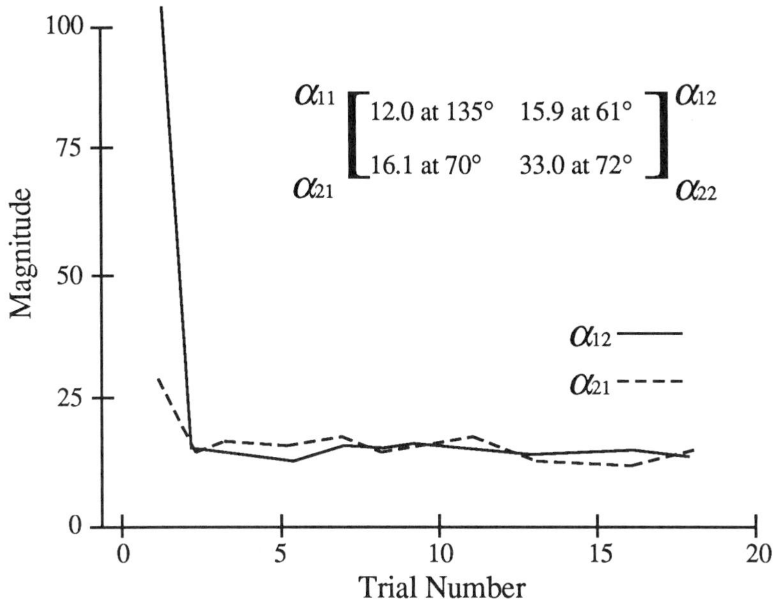


a) Magnitude of Influence Coefficients

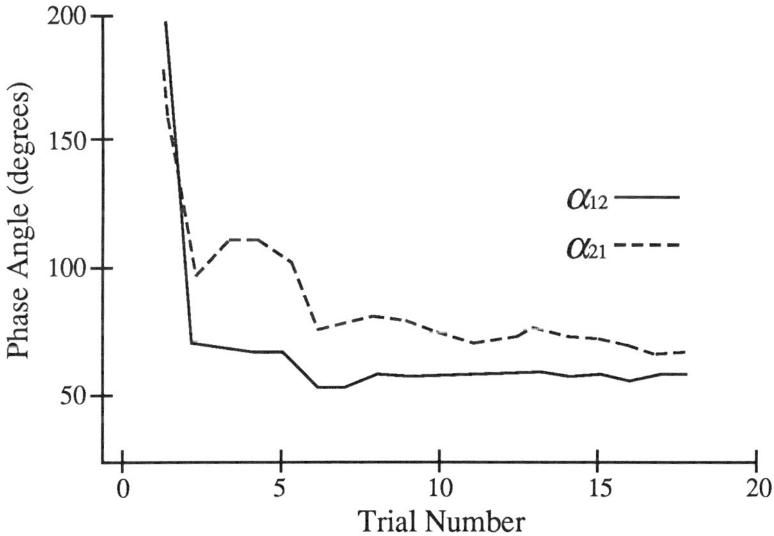


b) Phase Angles of Influence Coefficients

**Figure 9.4** Development of Symmetry in Measured Influence Coefficient Matrix for Demo Rig

The use of reciprocity with the Unified Balancing Approach can reduce the number of individual trial mass runs to a single run per mode. Unlike its use with influence coefficient balancing, a different plane may be used for each mode without sacrificing any of the benefits of reciprocity. The procedure for applying reciprocity to the Unified Balancing Approach is as follows.

1. After any particular mode (except for the highest mode of interest) has been satisfactorily balanced by the use of a modal correction mass set, a single trial mass is installed in a balancing plane which is at or near an anti-node for the mode most recently balanced. Vibration data is taken for all displacement probes and the corresponding influence coefficients are calculated.

2. By reciprocity, these influence coefficients are the same as those which relate the vibration data at the trial mass plane to unbalance masses located at the other balancing planes. It is only necessary to take into account differences in the calibration factors of the displacement probes and the radii at which the correction masses are applied, if different.

3. The appropriate influence coefficients can then be chosen for calculating the modal trial mass sets, as required.

Normally, the balancing of a rotor through $N$ critical speeds, using the Unified Balancing Approach, requires $\{[N(N+1)/2] - 1\}$ individual trial mass runs. However, using the reciprocity approach described above, the number of individual trial mass runs is reduced to $(N - 2)$; an individual trial mass run is not required for the first mode. Thus, a reduction of $\{[N(N-1)/2] + 1\}$ trial mass runs is obtained. So, if four modes must be balanced, the required number of individual trial mass runs can be reduced from nine to two. For the same example, the total number of trial mass runs is reduced from thirteen to six, which is a reduction of more than 50 percent.

# Glossary of Terms for High-Speed Rotor Balancing

*Balancing Mass Plane* (*Balancing Plane*) - Rotor axial position where balancing mass corrections may be applied.

*Check Balance* - Data taken at the balancing speed after the addition of a correction mass set. If the improvement in the balance is not sufficient, an additional correction mass set may be calculated using the check balance data. When using the Unified Balancing Approach, new modal influence coefficients may be generated with this data a used to calculate an additional modal correction mass set.

*Correction Mass* - A mass which is added to, or removed from, a rotor at a specified axial, radial and angular position; intended to modify the location of the center of mass of the rotor in such a way as to reduce the rotor's unbalance response.

*Critical Speed* - Synchronous natural frequency of a rotor.

*Dynamic Runout* - That part of the total runout that is truly a function of unbalance; determined by subtracting the static runout from the total runout.

*Field Balancing* - In-situ balancing of a rotor (rigid or flexible) which is fully installed and operating.

*Flexible* (*High-Speed*) *Rotor Balancing* - The balancing of rotors which operate in the vicinity of, or above, one or more flexural critical speeds. Thus, the shape of the rotor, and consequently of the mass centroidal axis, changes with speed.

*Flexural Critical Speed* - A critical speed at which the rotor exhibits significant flexural deformation.

*Individual Trial Mass* - A single trial mass used to generate influence coefficients for individual balance mass planes; used for influence coefficient balancing, or for calculating modal trial mass set ratios when using the Unified Balancing Approach.

*Influence Coefficient* - A complex value representing the effect of the addition of a unit trial mass in a specific balancing plane on the rotor response at a particular measurement plane.

*Influence Coefficient Balancing* - An entirely empirical, flexible rotor balancing method which uses known trial masses to experimentally determine the sensitivity of a rotor; and subsequently uses this sensitivity information to determine a set of discrete correction masses that will minimize synchronous vibrational amplitudes.

*Measurement Error* - Random or biased variations in empirical vibration measurements which are due to mechanical and/or electrical noise and other uncontrollable, variable physical parameters.

*Measurement Plane* - Rotor axial position where unbalance response measurements may be made.

*Modal Balancing* - Any of a group of flexible rotor balancing methods which assume orthogonal modes of vibrational deformation and strive to compensate for them individually.

*Modal Correction Mass Set* - A group of correction masses used to balance a particular mode while having no effect on other particular modes. The number of masses (and balancing planes) in this set is greater than or equal to the mode number of the mode being balanced. The magnitude ratios and relative angles of the masses in the set are the same as for the modal trial mass set used to excite the same mode, but the actual magnitudes and angles are calculated using rotor response data from the modal trial mass runs.

*Modal Influence Coefficient* - A complex value representing the effect of the addition of a unit modal trial mass set on the rotor response at a particular measurement plane.

*Modal Trial Mass Set* - A group of trial masses used to excite a particular mode while having no effect on other particular modes. The number of masses (and balancing planes) in this set is greater than or equal to the mode number of the mode being excited. The magnitude ratios and relative angles of the masses in the set are calculated using rotor response data from individual mass runs, but an arbitrary complex multiplier can be applied to the set.

*Mode Number* - Generally, the order in which critical speeds occur when the speed of the rotor-bearing system is increased. For generally rigid bearings, this number refers to the number of anti-nodes in the mode shape. For generally flexible bearings with distinctly separate rigid body modes, this number will be one for the first rigid body mode and one greater than the number of rigid body modes (not including additional modes due to bearing anisotropy) for the first flexural mode. When balancing the $n^{th}$ mode, there are $n-1$ modes with lower critical speeds (and possibly some modes with higher critical speeds), at which the vibration level must not be aggravated. Critically damped modes are generally not counted.

*Mode Shape* - Shape of the response of a rotor-bearing system at a critical speed.

*Orthogonality* (defined mathematically in Chapter 5) - Property whereby a group of functions, such as mode shapes or unbalance distributions, are mutually independent and are related to another group of functions in a one-to-one fashion.

*Phase Angle* - A measurement used to identify the circumferential location of a trial or correction mass, or unbalance response peak amplitude, relative to a specific angular reference position on the rotor.

*Reciprocity* - A property of linear static or dynamic response whereby two points in a structure exhibit identical sensitivity to one another, regardless of which is used as the forcing point and which as the response point.

*Residual (Unbalance) Response* - The unbalance response of a rotor remaining after the completion of one or more balancing operations.

*Rotor Balancing* - The application of any of a variety of methods for reducing the unbalance response of a rotor through the addition of correction masses to adjust the rotor's mass centroidal axis.

*Static Runout* - That part of the total runout that is due to synchronous structural and electrical sources, and does not change with rotational speed (and is thus not a function of unbalance); generally significant only for displacement sensors.

*Supercritical Rotor* - A rotor that operates above one or more critical speeds.

*Synchronous Response* - Rotor whirl whose frequency is identical to its rotational frequency.

*Total Runout* - The total synchronous component of the measured signal from a vibration sensor.

*Trial Mass* - A single mass, or group of masses, of known magnitude installed in a rotor at a specific balancing plane and angle, to generate data used for calculating influence coefficients of the rotor-bearing system for this balancing mass location (or combination) at speeds and measurement planes at which data are taken.

*Two Trial Masses Per Plane* (alternatively referred to as 0° and 180° trial masses) - Trial mass data taken with a trial mass (or set of masses) located at the specified trial mass angle (or angles) and then moved to 180 degrees from that position (or those positions). The additional data is used to calculate the measurement error inherent in the test data and to correct for that error.

*Unbalance* - The generally accepted term for the eccentricity of the center of mass of a rotor relative to its center of rotation (although the proper term is *imbalance*).

*Unbalance Response* - Synchronous response of a rotor due to a lack of concentricity of its mass centroidal axis, with respect to the center of rotation.

*Uncorrected Rotor* - Rotor condition at the beginning of a balancing step, with no trial or correction masses installed.

*Unified Balancing Approach* (UBA) - A flexible rotor balancing method which utilizes the empirical nature of influence coefficient balancing, while taking advantage of the modal behavior of the rotor.

# References


1. Avallone, Eugene A., and Baumeister, Theodore III, editors, *Mark's Standard Handbook for Mechanical Engineers*, ninth edition, McGraw-Hill Book Company, New York, 1987.

2. Badgley, R.H., "Implications of Multiplane-Multispeed Balancing for Future Turbine Engine Design and Cost," paper presented at the 1975 Aerospace Engineering & Manufacturing Meeting of the Society of Automotive Engineers, San Diego, California, (September 30-October 4) 1974.

3. Badgley, R.H., "Modern Influence Coefficient Techniques for Multiplane Rotor Balancing in the Factory, Test Cell and Field," *Vibrations in Rotating Machinery,* Institute of Mechanical Engineers' Conference Publications 1976-9, pp. 201-207, 1976.

4. Badgley, R.H., "The Potential Impact of Multiplane-Multispeed Balancing on Gas Turbine Production and Overhaul Costs," American Society of Mechanical Engineers Paper No. 74-GT-94, presented at the American Society of Mechanical Engineers Gas Turbine Conference and Products Show, Zurich, Switzerland, (April) 1974.

5. Badgley, R.H., "Recent Developments in Multiplane-Multispeed Balancing of Flexible Rotors in the United States," presented at the Symposium on Dynamics of Rotors, IUTAM, Lyngby, Denmark, (August 12) 1974.

6. Badgley, R.H., and Rieger, N.F., "The Effects of Multiplane Balancing on Flexible Rotor Whirl Amplitudes," Society of Automotive Engineers Paper No. 730102, presented at Society of Automotive Engineers International Automotive Engineering Congress, Detroit, Michigan, (January 8-12) 1973.

7. Badgley, R.H., Smalley, A.J., and Fleming, D.P., "Drive-Train Dynamics Technology: State-of-the-Art and Design of a Test Facility for Advanced Development," American Society of Mechanical Engineers Paper No. 75-DET-74, (September) 1975.

8. Badgley, R.H., Smalley, A.J., Malanoski, Stanley, Hamm, Robert, and Winn, Leo W., "Rotor Balancing," *Machinery Vibration Seminar Course Notes,* Mechanical Technology Incorporated, 1973.

9. Badgley, R.H., and Tessarzik, J.M., "Balancing of High-Speed Interconnect Shafting for Operation above Multiple Bending Critical Speeds," American Helicopter Society Paper No. 873, presented at the 30th Annual National Forum of the American Helicopter Society, Washington, D.C., (May) 1974.

10. Baier, R., and Mack, J., *Design and Test Evaluation of a Supercritical Speed Shaft,* The Boeing Company, Vertol Division, Morton, Pennsylvania, USAAVLABS Technical Report 66-49/R458, (June) 1968.

11. Baker, J.G., "Methods of Rotor-Unbalance Determination," *Transactions of the American Society of Mechanical Engineers, Journal of Applied Mechanics*, Vol. 61, pp. A1 - A6, (March) 1939.

12. Bigret, Roland, Curami, Andrea, Frigeri, Claudio, and Macchi, Nagelo, "Use of In-Field Computer for Balancing High Power Turbo-Machinery," American Society of Mechanical Engineers Paper No. 77-DET-11, (September) 1977.

13. Bishop, R.E.D., "On the Possibility of Balancing Rotating Flexible Shafts," *Journal of Mechanical Engineering Science*, Vol. 24, No. 4, pp. 215-220, 1982.

14. Bishop, R.E.D., "The Vibrations of Rotating Shafts," *Journal of Mechanical Engineering Science*, Vol. 1, No. 1, p. 50, 1959.

15. Bishop, R.E.D., and Gladwell, G.M.L., "The Vibration and Balancing of an Unbalanced Flexible Rotor," *Journal of Mechanical Engineering Science*, Vol. 1, No. 1, p. 66, 1959.

16. Bishop, R.E.D., and Mahalingam, S., "Some Experiments in the Vibration of a Rotating Shaft," *Proceedings of the Royal Society of London*, Series A, Vol. 292, p. 537, 1966.

17. Bishop, R.E.D., and Parkinson, A.G., "On the Isolation of Modes in the Balancing of Flexible Shafts," *Proceedings of the Institute of Mechanical Engineers*, Vol. 177, No. 16, p. 407, 1963.

18. Bishop, R.E.D., and Parkinson, A.G., "On the Use of Balancing Machines for Flexible Rotors," *Transactions of the American Society of Mechanical Engineers, Journal of Engineering for Industry*, Vol. 94, p. 561, 1972.

19. Bishop, R.E.D., and Parkinson, A.G., "Second-Order Vibration of Flexible Shafts," *Phil. Transactions of the Royal Society of London*, Series A, Vol. 259, p. 1, 1965.

20. Black, H.F., and Nuttal, S.M., "Modal Resolution and Balancing of Synchronous Vibrations in Flexible Rotors with Non-Conservative Cross Coupling," *Proceedings of Conference on Vibrations in Rotating Machinery*, Cambridge, England, pp. 151-158, 1976.

21. Bulanowski, E.A., "Practical Considerations for a Rated Speed Shop Balance," Paper. No. F-30-4-14-G, *Proceedings of the Seventh Turbomachinery Symposium*, pp. 87-94, 1976.

22. Church, A.H., and Plunkett, R., "Balancing Flexible Rotors," *Transactions of the American Society of Mechanical Engineers, Journal of Engineering for Industry*, Vol. 83, No. 4, pp. 383-389, (November) 1961.

23. Cundiff, R.A., Badgley, R.H., and Reddecliff, J., "Design, Manufacture, and Operation of a Dynamic Simulator for Testing Advanced Small Gas Turbine Engine Components," paper presented at the Symposium on Propulsion System Structural Integration and Engine Integrity, Monterey, California, (September 6) 1974.

24. Cundiff, R.A., Badgley, R.H., and Reddecliff, J., *Pneumo-mechanical Critical Speed Control for Gas Turbine Engine Shafts,* Paper No. AFPL-TR-73-101, (November) 1973.

25. Damon, W.E., *Research on Laser Balancing of Rotating Turbine Components,* Air Force Materials Laboratory Report No. AFML-TR-69-206, 1969.

26. Darlow, Mark S., "Balancing of High-Speed Machinery: Theory, Methods and Experimental Results," *Mechanical Systems and Signal Processing* Vol. 1, No. 1, pp. 105-134, 1987.

27. Darlow, M.S., "The Identification and Elimination of Non-Independent Balance Planes in Influence Coefficient Balancing," American Society of Mechanical Engineers Paper No. 82-GT-269 presented at the American Society of Mechanical Engineers International Gas Turbine Conference, (April) 1982

28. Darlow, M.S., "In Situ Balancing of Flexible Rotors Using Influence Coefficient Balancing and the Unified Balancing Approach," American Society of Mechanical Engineers Paper No. 83-GT-178, presented at the American Society of Mechanical Engineers International Gas Turbine Conference, (March) 1983.

29. Darlow, M.S., *A Unified Approach to the Mass Balancing of Rotating Flexible Shafts*, Ph.D. dissertation, University of Florida, (June) 1980.

30. Darlow, Mark, and Fanuele, Frank, *User's Manual for MTI Balancing Programs,* Mechanical Technology Incorporated Computer Program User's Manual, 1977.

31. Darlow, M.S., and Smalley, A.J., "Application of the Principle of Reciprocity to Flexible Rotor Balancing," *Transactions of the American Society of Mechanical Engineers, Journal of Mechanical Design*, Vol. 104, No. 2, pp. 329-333, (April) 1982.

32. Darlow, M.S., and Smalley, A.J., *Design and Test of a Scale Model Test Rig for Supercritical Power Transmission Shafting,* Mechanical Technology Incorporated Technical Report No. 78TR41, NASA Contractor Report CR-3155, prepared for NASA-Lewis Research Center, under Contract No. NAS3-16824, (September) 1978.

33. Darlow, M.S., and Smalley, A.J., "Design and Test of a Squeeze-Film Damper for a Flexible Power Transmission Shaft," *Topics in Fluid Film Bearing and Rotor Bearing System Design and Optimization*, American Society of Mechanical Engineers, pp. 43-51, (April) 1978.

34. Darlow, M.S., Smalley, A.J., and Fleming, D.P., "Demonstration of a Supercritical Power Transmission Shaft," published in the *Proceedings of the National Conference on Power Transmission,* Chicago, Illinois, (November 7-9) 1978.

35. Darlow, M.S., Smalley, A.J., and Parkinson, A.G., "Demonstration of a Unified Approach to the Balancing of Flexible Rotors," American Society of Mechanical Engineers Paper No. 80-GT-87, *Transactions of the American Society of Mechanical Engineers, Journal of Engineering for Power,* 1980.

36. Darlow, M.S., Smalley, A.J., and Parkinson, A.G., "A Unified Approach to Flexible Rotor Balancing: Outline and Experimental Verification," presented at the Institute of Mechanical Engineers' Conference on Vibrations in Rotating Machinery, Cambridge, England, (September) 1980.

37. DeMuth, R.S., Fleming, D.P., and Rio, R.A., "Laser Balancing Demonstration on a High-Speed Flexible Rotor," American Society of Mechanical Engineers Paper No. 79-GT-56, (March) 1979.

38. Den Hartog, J.P., *Mechanical Vibrations*, McGraw-Hill Book Company, Inc., New York, 1934. (fourth edition was published in 1956)

39. Dimentberg, F.M., "Theory of Balancing Flexible Rotors," *Russian Engineering Journal*, Issue No. 11, 1964.

40. Dimentberg, F.M., *Vibrations of Rotating Shafts*, Butterworth and Co., Ltd., London, 1961.

41. Dobson, C.P., *Laser Balancing of Verge and Shaft Assemblies,* Bendix Corporation Report BDX-613-724, prepared under USAEC Contract AT(29-1) 613, 1969.

42. Drechsler, J., "A Combination of Modal Balancing and the Influence Coefficient Method," *Proceedings of the Fourth World Congress on the Theory of Machines and Mechanisms,* Newcastle/Tyne, 1975.

43. Drechsler, J., "Systematic Combination of Experiments and Data Processing in Balancing of Flexible Rotors," *Vibrations in Rotating Machinery,* Institute of Mechanical Engineers' Conference Publications 1976-9, pp. 129-132.

44. El-hadi, Ibrahim, "Zusammenstellung, kritische Untersuchung und Weiterenlwicklung der Verfahren zum Auswuchten betriebsmabig aufgestellter Maschinen mit starren und mit elastischen Laufern."

45. Euler, J.A., and Speckhart, F.H., "Rotor Balancing Without Phase Measurement," American Society of Mechanical Engineers Paper No. 85-DET-117, Presented at the American Society of Mechanical Engineers Design Engineering Division Conference and Exhibit on Mechanical Vibration and Noise, Cincinnati, Ohio, (September 10-13) 1985.

46. Everett, Louis J., "Two Plane Balancing of a Rotor System Without Phase Measurements," *Journal of Vibration, Acoustics and Reliability Stress in Design,* Vol. 109, No. 2, pp. 162-167, (April) 1987.

47. Fawzy, I., and Bishop, R.E.D., "A Strategy for Investigating the Linear Dynamics of a Rotor in Bearings," *Vibrations in Rotating Machinery*, Institute of Mechanical Engineers' Conference Publications 1976-9, pp. 239-244.

48. Federn, I.K., "Looking to the Future in Balancing," presented at the Avery Symposium on Dynamic Balancing, (March) 1964.

49. Fujisawa, F., Siobato, K., Sato, K., Imai, T., and Shoyama, E., "Experimental Investigation of Multi-Span Rotor Balancing Using Least Squares Method," American Society of Mechanical Engineers Paper No. 79-WA/DE-1,.*Transactions of the American Society of Mechanical Engineers, Journal of Mechanical Design,* 1980.

50. Gasch, R. and Dreschlef, J., "Modales Auswuchten Elastischer Lau fur ohne Testgewichts-Setzungen," VDI-Berichte, Vol. 320, pp. 45-54, 1978.

51. Giers, A., "Practice of Flexible Rotor Balancing," *Vibrations in Rotating Machinery*, Institute of Mechanical Engineers' Conference Publications 1976-9, pp. 33-42.

52. Giordano, J., and Zorzi, E., *HPOTP Low-Speed Flexible Rotor Balancing-Phase I, Final Report*, Mechanical Technology Incorporated technical report 85TR28 for NASA Marshall Space Flight Center, 1985.

53. Gladwell, G.M.L., and Bishop, R.E.D., "The Vibration of Rotating Shafts Supported in Flexible Bearings," *Journal of Mechanical Engineering Science*, Vol. 1, No. 1, p. 66, 1959.

54. Gleizer, A.I., "Evaluating the Accuracy of Flexible Rotor Balancing on High-Speed Balancing Stands," *Izvestiya VUZ. Aviatsionnaya Tekhnika,* UDC 534, Vol. 22, No. 3, pp. 78-80, 1979.

55. Gleizer, A.I. and Bulychev, V.A., "Calculation of Effectiveness of Flexible Compressor Rotor Balancing on Low-Speed Balancing Machines," *Izvestiya VUZ. Aviatsionnaya Tekhnika*, Vol. 29, No. 1, pp. 16-20, 1986.

56. Gleizer, A.I., "Optimizing Gas Turbine Engine Flexible Rotor Balancing by the LP-Search Method," Izvestiya VUZ, *Aviatsionnaya Tekhnika,* Vol. 21, No. 1, pp. 111-113, 1978.

57. Gnielka, P., "Modal Balancing of Flexible Rotors Without Test Runs: An Experimental Investigation," *Journal of Sound and Vibration*, Vol. 90, No. 2,. pp. 157-172, 1983.

58. Gosiewski, Z., "Automatic Balancing of Flexible Rotors, Part I: Theoretical Background," *Journal of Sound and Vibration,* Vol. 100, No. 4, pp. 551-567, 1985.

59. Gosiewski, Z., "Automatic Balancing of Flexible Rotors, Part II, Synthesis of System," *Journal of Sound and Vibration* , Vol. 114, No. 1, pp. 103-119, 1987.

60. Goodman, T.P., "A Least-Squares Method for Computing Balance Corrections," *Transactions of the American Society of Mechanical Engineers, Journal of Engineering for Industry*, Vol. 86, No. 3, pp. 273-279, (August) 1964. Includes discussion by Heymann.

61. Grobel, L.P., "Balancing Turbine-Generator Rotors," *General Electric Review*, Vol. 56, No. 4, p. 22, 1953.

62. Harris, C.M., and Crede, C., editors, *Shock and Vibration Handbook*, first edition, McGraw-Hill, New York, 1961.

63. Harris, C.M., and Crede, C., editors, *Shock and Vibration Handbook*, second edition, McGraw-Hill, New York, 1976.

64. Hogan, B.J., "Compact System Performs Multiplane Rotor Balancing," *Design News*, Vol. 5, No. 19, pp. 92-94, 1986.

65. Hundal, M.S., and Harker, R.J., "Balancing of Flexible Rotors Having Arbitrary Mass and Stiffness Distribution," *Transactions of the American Society of Mechanical Engineers, Journal of Engineering for Industry*, Vol. 88, No. 2, pp. 217-223, (May) 1966.

66. Iwatsubo, T., "The Balancing of Flexible Rotors," (Theoretical Relation of Each Balancing Method and Accuracies), *Bulletin of the Japanese Society of Mechanical Engineers*, Vol. 23, No. 186, Paper No. 186-22, pp. 2096-2103, (December) 1980.

67. Jackson, Charlie, "Using the Orbit [Lissajous] to Balance Rotating Equipment," American Society of Mechanical Engineers Paper No. 70-Pet-30, (September) 1970.

68. Jeffcott, H.H., "The Lateral Vibration of Loaded Shafts in the Neighborhood of a Whirling Speed--The Effect of Want of Balance," *Phil. Mag.*, Series 6, Vol. 37, pp. 304, 1919.

69. Kellenberger, W., "Balancing Flexible Rotors on Two Generally Flexible Bearings," *Brown Boveri Review*, Vol. 54, No. 9, pp. 603-617, 1967.

70. Kellenberger, W., Weber, H., and Meyer-Baden, H., "Overspeed Test Facilities of the Group - Overspeed Testing and Balancing of Large Rotors," *Brown Boveri Review*, Vol. 6, No. 76, pp. 399-411.

71. Kellenberger, W., "Should a Flexible Rotor Be Balanced in N or (N2) Planes?" *Journal of Engineering for Industry*, Vol. 94, p. 548, 1972.

72. Kennedy, C.C., and Pancu, D.D.P., "Use of Vectors in Vibration Measurements and Analysis," *Journal of Aeronautical Science*, Vol. 14, 1947.

73. Korn, G.A., and Korn, T.M., *Mathematical Handbook for Scientists and Engineers,* second edition, McGraw-Hill Book Company, New York, 1968.

74. Kroon, R.P., "Balancing of Rotating Apparatus--I," *Transactions of the American Society of Mechanical Engineers, Journal of Applied Mechanics*, Vol. 10, p. A-225, 1943.

75. Kroon, R.P., "Balancing of Rotating Apparatus--II," *Transactions of the American Society of Mechanical Engineers, Journal of Applied Mechanics*, Vol. 11, pp. A-47-A-50, 1944.

76. Kushue, M. Ya., and Shlyaktin, A.V., "Modal Approach to Balancing with Additional Constraints," *Isvestiya Akademii Nauk SSR, Mekhanikai mashinostroyeniye*, No. 2, 1966.

77. Larsson, Lars-Ove, "On the Determination of the Influence Coefficients in Rotor Balancing, Using Linear Regression Analysis," *Vibrations in Rotating Machinery,* Institute of Mechanical Engineers' Conference Publications 1976-9, pp. 93-97.

78. Lee, T.W., and Kakad, Y.P., "Heuristic Combinatorial Optimization in the Balancing of Flexible Rotors," unpublished manuscript, 1982

79. LeGrow, J.V., "Multiplane Balancing of Flexible Rotors--A Method of Calculating Correction Weights," American Society of Mechanical Engineers Third Vibration Conference Paper No. 71-Vibr-52, Toronto, Canada, (September) 1971.

80. Lindley, A.G., and Bishop, R.E.D., "Some Recent Research of the Balancing of Large Flexible Rotors," *Proceedings of the Institute of Mechanical Engineers*, Vol. 177, No. 30, p. 811, 1963.

81. Little, R.M., "The Application of Linear Programming Techniques to Balancing Flexible Rotors," Ph.D. Thesis, University of Virginia, 1971.

82. Little, R.M., and Pilkey, W.D., "A Linear Programming Approach for Balancing Flexible Rotors," *Transactions of the American Society of Mechanical Engineers, Journal of Engineering for Industry,* Vol. 98, No. 3, pp. 1030-1035, 1976.

83. Loewy, R.G., and Piarulli, V.J., *Dynamics of Rotating Shafts*, Shock and Vibration Monograph, No. 4, Naval Research Laboratory, Washington, D.C., 1969.

84. Lund, J.W., "A Method for Using the Free Shaft Modes in Rotor Balancing," *Vibrations in Rotating Machinery*, Institute of Mechanical Engineers' Conference Publications 1976-9, pp. 65-71.

85. Lund, J.W., "Elliptical Orbit Definitions," Appendix to *Unbalance Response of a Flexible Rotor*, Mechanical Technology Incorporated Computer Program Users Manual, n.d.

86. Lund, J.W., personal communication, 1979.

87. Lund, J.W., and Tonneson, J., "Analysis and Experiments in Multiplane Balancing of Flexible Rotors," *Transactions of the American Society of Mechanical Engineers, Journal of Engineering for Industry*, Vol. 94, No. 1, p. 233, (February) 1972.

88. Lund, J.W., and Tonneson, J., *Experimental and Analytic Investigation of High-Speed Rotor Balancing--Phase I*, Department of Machine Design, Technical University of Denmark, Research Report No. FR-8, Project No. FP-4, (December) 1970.

89. *The Scientific Papers of James Clerk Maxwell*, Dover, New York, 1890.

90. Maxwell, A.S., and Sanderson, A.F.P., "Site Balancing of a Large Flexible Rotor Containing Unbalance Eccentricity and Permanent Residual Bow," *Proceedings of the Fifth Turbomachinery Symposium*, pp. 23-30, 1974.

91. Meldal, A., "Auswuchten Elastischer rotoren," *Z. angew. Mach und Mech.*, Vol. 34, 1954.

92. Miwa, Shuzo, "Balancing of a Flexible Rotor," *Balancing of a Flexible Rotor 3rd Report*, Vol. 16, No. 100, 621-755 : 621-25, pp. 1562-1570, (October) 1973.

93. Moore, L.S., and Dodd, E.G., "Mass Balancing of Large Flexible Rotors," *G.E.C. Journal*, Vol. 31, No. 2, 1964.

94. Morton, P.G., "Modal Balancing of Flexible Shafts Without Trial Weights," *Proceedings of the Institute of Mechanical Engineers*, Vol. 199, No. C1, pp. 71-78, 1985.

95. Muster, D., and Flores, B., "Balancing Criteria and Their Relationship to Current American Practice," American Society of Mechanical Engineers Vibrations Conference Paper 69-VIBR-60, Philadelphia, 1969.

96. Nakai, Takafumi and Miwa, Shuzo, "Balancing of a Flexible Rotor," *Bulletin of the Japanese Society of Mechanical Engineers*, Vol. 23, No. 176, pp. 280-285, 1980.

97. Nicholas, J.C., Gunter, E.J., and Allaiire, P.E., "Effect of Residual Shaft Bow on Unbalance Response and Balancing of a Single Mass Flexible Rotor. Part I: Unbalance Response," *Transactions of the American Society of Mechanical Engineers, Journal of Engineering for Power,* Vol. 98, No. 2, pp. 171-181, (April) 1976.

98. Nicholas, J.C., Gunter, E.J., and Allaire, P.E., "Effect of Residual Shaft Bow on the Unbalance Response and Balancing of a Single Mass Flexible Rotor. Part II: Balancing," *Transactions of the American Society of Mechanical Engineers, Journal of Engineering for Power,* Vol. 98, No. 2, pp. 183-188, (April) 1976.

99. Noble, Ben, and Daniel, James W., *Applied Linear Algebra,* second edition, Prentice-Hall, Inc., Englewood Cliffs, New Jersey, pp. 138-143, 1977.

100. Parkinson, A.G., "An Introduction to the Vibration of Rotating Flexible Shafts," *Bulletin of Mechanical Engineering Education*, Vol. 6, p. 47, 1967.

101. Parkinson, A.G., "The Modal Interpretation of the Vibration of a Damped Rotating Shaft," *Vibrations in Rotating Machinery*, Institute of Mechanical Engineers Conference Publications 1976-9, pp. 263-269.

102. Parkinson, A.G., "On the Balancing of Shafts with Axial Asymmetry," *Proceedings of the Royal Society of London*, Series A, Vol. 292, p. 66, 1966.

103. Parkinson, A.G., "The Vibration and Balancing of Shafts Rotating in Asymmetric Bearings," *Journal of Sound and Vibration*, Vol. 2, No. 4, p. 477, 1965.

104. Parkinson, A.G., and Bishop, R.E.D., "Residual Vibration in Modal Balancing," *Journal of Mechanical Engineering Science*, Vol. 7, No. 1, p. 33, 1965.

105. Parkinson, A.G., Darlow, M.S., Smalley, A.J., and Badgley, R.H., "An Experimental Introduction to the Development of a Unified Approach to Flexible Rotor Balancing," presented at Machinery Vibrations IV (Vibration Institute), Cherry Hill, New Jersey, (November) 1980.

106. Parkinson, A.G., Darlow, M.S., Smalley, A.J., and Badgley, R.H., "An Introduction to a Unified Approach to Flexible Rotor Balancing," American Society of Mechanical Engineers Paper 79-GT-161 presented at the American Society of Mechanical Engineers Gas Turbine Conference, (March) 1979.

107. Parkinson, A.G., Darlow, M.S., Smalley, A.J., and Badgley, R.H., "A Theoretical Introduction to the Development of a Unified Approach to Flexible Rotor Balancing," *Journal of Sound and Vibration*, Vol. 68, No. 4, pp. 489-506, 1980.

108. Pilkey, W.D., and Bailey, J.D., "Constrained Balancing Techniques for Flexible Rotors," *Transaction of the American Society of Mechanical Engineers, Journal of Mechanical Design*, Vol. 101, No. 2, pp. 304-308, 1979.

109. Pilkey, W.D., Bailey, J., and Smith, P.D., "A Computational Technique for Optimizing Correction Weights and Axial Location of Balance Planes of Rotating Shafts," *Transactions of the American Society of Mechanical Engineers, Journal of Vibration, Acoustics, Stress and Reliability in Design*, Vol. 105, No. 1, January 1983, pp. 90-93.

110. Popick, H., and Roberts, D.L., "Laser Apparatus for Removing Material from Rotating Objects," U.S. Patent No. 3,472,998, 1969.

111. Rayleigh, J.W.S., *The Theory of Sound,* Vol. I, second edition, Dover Publications, pp. 150-160 (Sections 107-116), New York, 1894 (1945 printing)

112. Rieger, N.F., "Balancing High-Speed Rotors to Reduce Vibration Levels," Paper presented at American Society of Mechanical Engineers Design Conference, Chicago, (May 11) 1972.

113. Rieger, N.F., *Computer Program for Balancing of Flexible Rotors*, Mechanical Technology Incorporated Report 67TR68, (September) 1967.

114. Rieger, N.F., "Flexible Rotor-Bearing System Dynamics: III. Unbalance Response and Balancing of Flexible Rotors in Bearings," *American Society of Mechanical Engineers Monograph*, 1973.

115. Rieger, N.F., and Badgley, R.H., "Flexible Rotor Balancing of a High-Speed Gas Turbine Engine," Society of Automotive Engineers Paper No. 720741, presented at Society of Automotive Engineers National Combined Farm, Construction & Industrial Machinery and Power Plant Meeting, Milwaukee, Wisconsin, (September 11-14) 1972.

116. Rieger, N.F., and Crofoot, J.F., *Vibrations of Rotating Machinery, Part I: Rotor-Bearing Dynamics*, The Vibration Institute, Clarendon Hills, Illinois, 1977.

117. Saito, S., and Azawa, T., "Balancing of Flexible Rotors by the Complex Modal Method," *Transactions of the American Society of Mechanical Engineers, Journal of Vibration, Acoustics, Stress and Reliability in Design* , Vol. 105, No. 1, pp. 94-100, (January) 1983.

118. Shimada, K., Miwa, S., and Nakai, T., "Balancing of a Flexible Rotor," *The Fifth Report, Vibration and Balancing of a Flexible Rotor on Flexible Bearing with Viscous Damping*, Paper No. 151-7, pp. 44-50, 1975.

119. Shiohata, K., and Fujisawa, F., "Balancing Method of Multi-span, Multi-bearings Rotor System" (2nd Report: Balancing for a Flexible Rotor Considering Constraint of the Correction Mass), *Bulletin of the Japanese Society of Mechanical Engineers*, Vol. 23, No. 185, pp. 1894-1898, (November) 1980.

120. Shiohata, K., Fujisawa, F., and Sato, K., "Method of Determining Locations of Unbalances in Rotating Machines," *Transactions of the American Society of Mechanical Engineers, Journal of Mechanical Design*, Vol. 104, No. 2 pp. 290-295, (April) 1982.

121. Smalley, A.J., Tessarzik, J.M., and Badgley, R.H., "The Stability of an Asymmetric Rotor in Damped Supports," American Society of Mechanical Engineers Paper No. 78-GT-172, (April) 1978.

122. Smalley, A.J., Tessarzik, J.M., and Rio, R.A., *Balancing Techniques for High-Speed Flexible Rotors,*" Mechanical Technology Incorporated Technical Report No. MTI-77TR2, NASA Contractor Report CR-2975, prepared for NASA-Lewis Research Center under Contract No. NAS3-18520, (April) 1978.

123. Stadelbauer, D.G., "Balancing Machines Reviewed," Schenk Trebel Corporation, n.d.

124. Tang, T.M., and Trumpler, P.R., "Dynamics of Synchronous-Precessing Turborotors with Particular Reference to Balancing, Part II--Application," *Transactions of the American Society of Mechanical Engineers, Journal of Applied Mechanics*, Vol. 90, pp. 25-30, 1968.

125. Tessarzik, J.M., *Flexible Rotor Balancing by the Exact Point-Speed Influence Coefficient Method*, Mechanical Technology Incorporated Technical Report No. MTI-70TR59, NASA Contractor Report CR-72774, prepared for NASA-Lewis Research Center under Contract No. NAS3-13473, (October) 1970.

126. Tessarzik, J.M., *Flexible Rotor Balancing by the Influence Coefficient Method--Multiple Critical speeds with Rigid or Flexible Supports,* Mechanical Technology Incorporated Technical Report No. MTI-75TR3, NASA Contractor Report CR-2553, prepared for NASA-Lewis Research Center under Contract No. NAS3-14420, (February) 1975.

127. Tessarzik, J.M., *Flexible Rotor Balancing by the Influence Coefficient Method: Part 1--Evaluation of the Exact Point-Speed and Least-Squares Procedures*, Mechanical Technology Incorporated Technical Report No. MTI-72TR32, NASA Contractor Report CR-121107, prepared for NASA-Lewis Research Center under Contract No. NAS3-14420, (July) 1972.

128. Tessarzik, J.M., and Badgley, R.H., "Experimental Evaluation of the Exact Point-Speed and Least-Squares Procedures for Flexible Rotor Balancing by the Influence Coefficient Method," *Transactions of the American Society of Mechanical Engineers, Journal of Engineering for Industry*, Vol. 96, No. 2, pp. 633-643, (May) 1974.

129. Tessarzik, J.M., Badgley, R.H., and Anderson, W.J., "Flexible Rotor Balancing by the Exact Point-Speed Influence Coefficient Method," *Transactions of the American Society of Mechanical Engineers, Journal of Engineering for Industry*, Vol. 94, No. 1, pp. 148-158, (February) 1972.

130. Tessarzik, J.M., Badgley, R.H., and Fleming, D.P., "Experimental Evaluation of Multiplane-Multispeed Rotor Balancing through Multiple Critical Speeds," *Transactions of the American Society of Mechanical Engineers, Journal of Engineering for Industry,* (August) 1976.

131. Tessarzik, J.M., Darlow, M.S., and Badgley, R.H., *Simultaneous Multiplane-Multispeed Balancing of 2 Steam Turbines for a Land-Based Submarine Drive Test Stand,* Mechanical Technology Incorporated Technical Report No. 76TR33, for NAVSEC (U.S. Navy), Philadelphia, Pennsylvania, (May) 1976.

132. Tessarzik, J.M., and Fleming, D.P., "Tests of Laser Metal Removal for Future Flexible Rotor Balancing in Engines," Society of Automotive Engineers Paper No. 750170, presented at Automotive Engineering Congress and Exposition, Detroit, Michigan, (February) 1975.

133. Thearle, E.L., "Dynamic Balancing of Rotating Machinery in the Field," *Transactions of the American Society of Mechanical Engineers, Journal of Applied Mechanics*, Vol. 56, pp. 745-753, 1934.

134. Thomson, W.T., *Theory of Vibration with Applications*, third edition, Prentice Hall, Inc., Englewood Cliffs, New Jersey, 1965.

135. Timoshenko, S., *Vibration Problems in Engineering*, D. Van Nostrand Co., Inc., Princeton, New Jersey, 1928. (fourth edition, with D.H. Young and W. Weaver Jr., was published by Wiley, New York, in 1974)

136. Tondl, A., *Some Problems in Rotordynamics*, Publishing House, Czechoslovakian Academy of Sciences, Prague, 1965.

137. Tonneson, J., *Experimental Investigation of High-Speed Rotor Balancing*, Department of Machine Design, Technical University of Denmark, Research Report No. FR-9, (December) 1972.

138. Tonneson, J., "Further Experiments on Balancing of a High-Speed Flexible Rotor," *Transactions of the American Society of Mechanical Engineers, Journal of Engineering for Industry*, Vol. 96, No. 2, pp. 431-440, (May) 1974.

139. Tse, Francis S., Morse, Ivan E., and Hinkle, Rolland T., *Mechanical Vibrations*, second edition, Allyn and Bacon, Inc., Boston, 1978.

140. Vance, John M., *Rotordynamics of Turbomachinery*, John Wiley and Sons, New York, 1988

141. Van de Vegte, J. and Lake, R.T., "Balancing of Rotating Systems During Operation," *Journal of Sound and Vibration* , Vol. 57, No. 2, pp. 225-235, 1978.

142. Wang, Jhy Horng, "A New Balancing Procedure for Multispan Flexible Rotor System," unpublished manuscript, 1986.

143. Woomer, E. and Pilkey, W., "The Balancing of Rotating Shafts by Quadratic Programming," *Journal of Mechanical Design*, Vol. 103, pp. 831-834, (October) 1981.

144. Yanabe, S., and Tamara, A., "Multi-Plane Balancing of Flexible Rotor Consisting of Two Disks," *Bulletin of the Japanese Society of Mechanical Engineers*, Vol. 12, No. 54, p. 1349, 1969.

# Index

## A



## B

## C



## D



## E

## F



## G



## H



## I



## J



## L